THE HISTORY
OF QUANTUM THEORY

THE FIFTH SOLVAY CONGRESS, BRUSSELS 1927
Reproduced with permission of the Solvay Institute
From left to right Top row: PICCARD, HENRIOT, EHRENFEST, HERZEN, DE DONDER, SCHROEDINGER, VERSCHAFFELT, PAULI, HEISENBERG, FOWLER, BRILLOUIN—Middle row: DEBYE, KNUDSEN, BRAGG, KRAMERS, DIRAC, COMPTON, DE BROGLIE, BORN, BOHR—Bottom row: LANGMUIR, PLANCK, CURIE, LORENTZ, EINSTEIN, LANGEVIN, GUYE, WILSON, RICHARDSON

THE HISTORY
OF QUANTUM THEORY

by

FRIEDRICH HUND
(formerly Professor at the University of Göttingen)

translated by

GORDON REECE, B.A.(Hons.), M.Sc.
Imperial College, London

BOOKS
10 East 53d St., New York 10022
(a division of Harper & Row Publishers, Inc.)

Published in the U.S.A. 1974 by
HARPER & ROW PUBLISHERS, INC.
BARNES & NOBLE IMPORT DIVISION

ISBN 06–493060–2

Made in Great Britain

FOREWORD

The rise of quantum theory in the years 1900 to 1927 is surely one of the major advances in the history of science—perhaps even one of the greatest intellectual advances ever made by mankind. Seventeenth century mechanics laid the foundations for an understanding of the motions of earth-bound and heavenly bodies. The development of electrodynamics in the nineteenth century made it possible to understand the phenomena of light, electricity and magnetism in terms of a single set of basic principles. Once quantum theory had been finalized, about forty years ago, it became possible to think in terms of a single fundamental science of simple natural phenomena, comprising occurrences at the atomic level, and thus unifying the theoretical foundations of physics and chemistry.

In this book, I have attempted to provide a survey of the history of quantum theory. We are now sufficiently far away in time to be able to assess the significance of each of the contributions—at least to some extent—and still sufficiently close to be able to draw on personal recollections, to be able to understand the pre-quantum situation, to understand some of the challenges of the experimental data—and even to be able to see the significance of some of the mistakes people made, and why they failed to spot them.

This book is intended for students, and even for those who may not have much knowledge of quantum theory, at least from the practical point of view. The outline of quantum theory given in the appendix, written as it is in roughly chronological order, should help them to follow the text. On the other hand, I hope that the book will also be of interest to physicists and physics teachers. It may well be that one understands physics all the better for knowing something of the difficulties that its discoverers had to overcome.

There is a fundamental and meticulous account of the history of quantum theory which is particularly thorough in dealing with the early history and subsequent explanation of the basic ideas: MAX JAMMER'S *The Conceptual Development of Quantum Mechanics*. Anyone who wishes to go more deeply into the history of

6

quantum theory will certainly want to read both JAMMER's book and the Historical Introduction to B. L. VAN DER WAERDEN's *Sources of Quantum Mechanics*.

I should like to thank Professor J. HAJDU of Cologne, and Dr H. KEITER for reading the manuscript and making constructive criticisms. I should also like to thank my wife, Dr I. HUND, and Dr KEITER for help in reading the proofs.

Friedrich Hund

Göttingen

CONTENTS

LIST OF ILLUSTRATIONS AND TABLES

10

1. SURVEY

Introduction

WHY do we study history? Not just to find out what happened—there must be a more specific reason. Perhaps it would be more accurate to say that we want to know how the present situation came about, in order to understand it better. This is precisely where the growth of an exact science differs from that of politics or sociology. From time to time science establishes a whole system of new theorems, which combine to form a closed field (by which we mean that it makes predictions that appear to be perfectly correct and would seem to have been thoroughly verified). Any new theory may well have to overcome established philosophical traditions. A good example of this process is given by the way the new mechanics of the seventeenth century superseded Aristotelian mechanics. The new theories in their turn gave rise to new philosophical traditions. It may well be that even modern science will one day have to make equally radical changes to accommodate those areas, such as quantum mechanics, where it is still incomplete. Studying the development of a science may help to improve our understanding of it. It is in this spirit that we shall study the history of quantum mechanics. Our aim will be a better understanding of the final version of quantum theory and thus of the whole of modern physics.

History is the story of how *a* possibility became *the* reality. The history of politics tells us how 'time' rejected countless possibilities and allowed just one to become a firm reality. In the history of science, on the other hand, 'time' has gradually caused truth to prevail. In the history of politics chance has therefore played an important role. That is not to say that it has played none in the history of science (PLANCK happened to see early results of precise measurements of the infra-red spectrum, BOHR happened to be working with RUTHERFORD in 1912). However, the fact is that in the history of science necessity has been more influential than chance. For example, of all the experimental indications of the existence of a fundamental quantum of action, it was the measurements of black-body radiation that at the time gave the most

information. Another crucial factor is of course the type of experimental techniques that happen to be available at any given
moment of time. Only from 1912 was it possible to measure the
wavelengths of X-rays, and it was not until 1925 that thin foil
was sufficiently developed to permit the measurement of the
interference of matter. The history of quantum theory is part
chance, part necessity and to a large extent a question of the
possibility of experimental verification. We shall need to focus
our attention on these three factors. Now and then we shall pause
to ask just how quantum mechanics might have developed if...
(if, say, PLANCK had decided to become a musician, or BOHR a
lawyer).

The history of physics is essentially that of a gradual realization
that all is not just as it seems—a departure from the 'perceptual'.
This involves a modification of the metaphysics that always
underlies the concepts of physicists, however reluctant they may
be to admit it. Aristotelian mechanics lay very close to the perceptual. In its later, more sophisticated form, the Aristotelian
concept of motion was roughly that the velocity of a moving
body was determined by the ratio of drive (force) to resistance.
This corresponds to the idea of dragging a load. NEWTON's version, on the other hand, in which the *rate of change* of velocity
was determined by the force, corresponds to the motion of a
particle in a vacuum. This departure from concrete, 'obvious'
reality did not make physics less accessible. On the contrary, at
each new stage in the process it became possible to understand
an increasing number of phenomena using a single approach, and
this included some phenomena that had previously been regarded
as belonging to quite different fields. However, one result of this
process was that much that had previously been treated as simple
was now recognized to be complex. Thus, for example, even
something as familiar as water becomes highly complicated when
looked at from the point of view of modern physics. Indeed it is
not yet fully understood.

The essential stages in this development were:

1. The ideal case of the particle left to its own devices, or pro-
   pelled by forces through empty space. The motion of the
   earth around the sun can be treated in this way to a large
   extent. (*Seventeenth century.*)
2. The introduction of the idea of the electromagnetic *field*,
   which gave a better understanding of electromagnetism and

light than the concept of the *forces* between electrically charged or magnetized bodies had ever managed to provide. (*Nineteenth century.*)

3. Quantum mechanics, which provided a theoretical treatment of the atom and derived chemistry from physical principles. (The first thirty years of the *Twentieth century.*)

4. Man begins to be able to handle forces of vastly different orders of magnitude, both in theory and in practice, starting with the atomic nucleus. (*The present.*)

This list shows the development of quantum theory as being bounded neatly in time. It actually belongs to the years 1900 to 1927, before Man began to be able to cope with other orders of magnitude. What preceded it we shall call classical physics.

Classical Physics

We may regard the mechanics of systems of particles as the core of classical physics. Here we assume the existence of forces between any two particles, acting along the line joining them and depending upon the distance between them (in the ideal case where there is no friction). We may therefore write 'equations of motion':

$$m_k \ddot{x}_k = F_k(x_{12}, x_{13} \ldots).$$

The suffix $k = 1, 2, \ldots$ refers to the particles; m_k denotes a mass, F_k a force-vector, x_k the position-co-ordinates, and the $x_{12}, \ldots$ the vector distances between them. In the nineteenth century this 'mechanics of forces' was quite happily derived from a mechanics based on certain 'principles'. Any such system could be characterized, for example, by a function called its 'Hamiltonian' $H(p_1, p_2, \ldots, q_1, q_2, \ldots)$. Even HELMHOLTZ still thought largely in terms of this system of particle mechanics. He saw it as one of the major tasks of physics to find the law that governed the forces between the atoms, which were regarded as particles. Those, indeed, were the lines along which most physicists were still thinking at the end of the nineteenth century.

In the course of the nineteenth century three major extensions were made to this core of classical physics, modifying and confirming the prevailing corpuscular-kinetic model of nature.

This is the point at which we must introduce the *theory of the electromagnetic field and of light*. First formulated by MAXWELL in about 1862, it was not until 1890 that the theory was put into a simple form and accepted by physicists. It was still open to question as far as HELMHOLTZ was concerned in 1888. It suggested a different general model of natural phenomena. Indeed, the question was sometimes raised as to whether it might perhaps be possible to explain all natural phenomena in terms of electromagnetic concepts. The next extension was *thermodynamics*, which has been on a sound theoretical basis since 1850. Thermodynamics clarified the concepts of heat and temperature and introduced the idea of entropy through the formula

$$dQ = TdS$$

for reversible phenomena (the increase in heat is equal to the temperature multiplied by the increase in entropy) and showed that the quantity S could not decrease in a closed system. The third major extension was that of *statistical physics*. It began with the kinetic theory of gases and reached its peak in BOLTZMANN's explanation of entropy

$$S \sim \ln W$$

in terms of the number, W, of events needed to represent a macroscopic state, given by the region of phase-space (q,p-space) which corresponded to the macrostate (1877). The great achievement of statistical physics was that of deriving thermodynamics from mechanical principles.

The overall picture of theoretical physics towards the end of the nineteenth century is thus characterized by two conceptual systems, that of mechanics and that of the electromagnetic field. Frequently the 'physics of matter', which used mechanics, was distinguished from the 'physics of the aether', where electromagnetic field theory held sway. There was a general feeling of optimism among physicists and, even more, among those who interpreted and popularized physics. It was felt that an essential general understanding of physical phenomena had already been achieved. But clouds could be seen hovering over this sunny picture.

At the dawn of the new century W. THOMSON, LORD KELVIN, gave a lecture entitled '19th-Century Clouds Over the Dynamical Theory of Heat and Light'.[1] It was customary in those days to think in material terms. It followed that physicists required a material carrier for the electromagnetic field quantities—they called it the aether. The field quantities described stresses of the aether, and the field equations even assigned it a transverse elasticity (shear elasticity), but they did not invest it with a volume elasticity (cf. a stiff jelly). It turned out to be impossible to deduce the velocity of the earth relative to the aether from the effect of the motion on the velocity of light. KELVIN saw the first of the 'clouds' in the questions: How can the heavenly bodies move without friction through the solid aether, and why is it impossible to deduce the velocity of the aether itself by experiment? The answer later given to the first question was that the aether was no longer considered to be material; it was therefore replaced by the concept of a 'field'. KELVIN's second question was answered by the theory of relativity. The second 'cloud' consisted in the fact that the measured specific heats of bodies were lower than the values given by the equipartition theorem. From statistical physics, as described above, it follows that every degree of freedom of a molecule in the medium must add $kT/2$ to the kinetic energy, and every harmonic oscillator likewise $kT/2$ to the potential energy. One mole of a diatomic gas, the molecules of which consist of two particles oscillating with respect to each other, should accordingly have energy $(6/2 + 1/2)RT$. The measured specific heat (at constant volume) was, however, not $(7/2)R$ but $(5/2)R$.

KELVIN's list included only those 'clouds' that lay over well-founded theories. He did not go into the question of the structure of the atom, a problem that was still barely understood.

The discrepancy mentioned above, between theory and experiment, in the case of specific heats, was used by J. W. GIBBS in the foreword to his *Statistical Mechanics* (1901) to underline the uncertain status of the assumption that had been made about the structure of matter.

The Novelty of Quantum Theory

The twentieth century has given physics the theory of relativity and a theory of gravitation, quantum theory and the theory of the atomic shell, further information about the nucleus and the beginnings of a theory of elementary particles. From our own point of view quantum theory would seem to be the most significant of these advances. It is significant in its power: it has solved the problems of matter and of the structure of the atom, and it

16 The History of Quantum Theory

has turned the fundamental concepts of chemistry into a part of physics. It is significant in its depth: it has enabled us to give a physical interpretation of the atom, by forcing a change in the very manner in which we describe nature.

Classical mechanics assumes, whether tacitly or explicitly, the 'axiom of certainty': to the variables, p_k and q_k, say, correspond magnitudes that must take precise values. Logically this assumption is tantamount to the strict determination of the future. Quantum theory, on the other hand, asserts a restriction on the extent to which we can determine physical quantities. It thus also implies a limited determination of the future. In the language of quantum mechanics, a 'simultaneous measurement of canonically conjugate variables p, q cannot determine these magnitudes more accurately than to within the uncertainties Δp, Δq given by $\Delta p \cdot \Delta q \approx h$, where h is the fundamental quantum of action'. In the general form of quantum theory in which p, q are mathematical entities more general than ordinary numbers, this situation is described more precisely by the equation

$$i(pq - qp) = \hbar = \frac{h}{2\pi}.$$

This involves a new kinematics to replace the classical mode of description, and a restriction on the extent to which the past determines the present and future.

It is possible to conceive of quantum theory in abstract terms as a new theory from which we can derive the classical theory by taking the limit $h \to 0$. But we can equally well obtain quantum theory by taking classical theory and modifying it. In the process of doing so, it transpires that we must modify our very ideas of particles. In the classical situation these ideas were intuitive*; they must be replaced by entirely non-intuitive concepts, so that we may ascribe wave properties even to a flow of matter. We must replace the classical, intuitive field or wave theory of matter by a non-intuitive one to be able to derive the idea of particles. Either of these approaches leads to the same quantum theory. A similar wave-particle duality holds in the case of light; but the quantum theory we need for this is a relativistic one and it is not complete to the same degree of sophistication as its non-relativistic counterpart. In the case of matter and light, the particle properties, momentum p and energy E, correspond to the wave properties:

* German: *anschaulich*.

wave number k (per 2π units of length) and frequency ω (per 2π units of time), as given by the equations

$$p = \hbar k \qquad E = \hbar \omega$$

In both the more abstract of these models and the one that is closer to classical concepts, the fundamental quantum of action h (or $\hbar = h/2\pi$) is the central quantity. *Quantum theory is the study of the role of* h *in nature.* To understand this role, physicists had to argue from the specific to the general. They recognized:

 h in statistics as a unit for counting 'events' (to determine how many events make up a macrostate) in the years 1900–13;
 h in the wave-particle duality for light, and thus as the quantity which governed the interaction of light and matter, 1905–23;
 h in atomic dynamics 1913–25;
 h in the wave-particle duality of matter 1923–6;
 h as a limitation of our ability to 'describe' in the classical sense, in the final version of quantum mechanics 1925–7.

This situation makes it easier to recount the historical development of quantum theory without separating those aspects that are closely related in either time or type. We shall accordingly consider: the history of quantum statistics (Chapter 2), the history of the quantum of light (3), the tentative form of atomic dynamics (4–7), and then the crises in atomic dynamics and their resolution (8–10), the matter-wave (11), and the ultimate completion of quantum mechanics (12–13) including its applications (14) and its extension (15).

There was a vast amount of experimental evidence for the various roles of h:

1. The application of statistical thermodynamics to the temperature-dependence of phenomena provides two pieces of evidence:

1a. At low temperatures, not all degrees of freedom necessarily contribute to the energy. The specific heats of gases at ordinary temperatures could be understood only if gas molecules were treated as rigid bodies. If two rigid spheres collide the degrees of freedom of rotation are unaffected, and there is therefore no procedure for obtaining the thermal equilibrium state of these degrees of freedom. Monatomic gases therefore have (at constant volume) the molar specific heat $3R/2$. When rigid rotating axisymmetric bodies collide, the rotation about the axis is not affected. Diatomic gases have a molar specific heat $5R/2$. This was the explanation given by BOLTZMANN. Of the rigid substances, the elements B, C (diamond), Al and Si gave a specific heat that was below the DULONG–PETIT value $3R$ per gram atom; the DULONG–PETIT value was the limit for high temperatures. Theories were adduced which took account of the difference between the oscillations in rigid bodies from those of harmonic oscillators. The decrease of specific heats at low temperatures could not be measured until 1910. Thus it was that some of the following results could only be obtained after the discovery of the quantum of action. The oscillation contribution is almost completely absent from the specific heat of gases. The contribution from the rotation decreases monotonically towards zero with the temperature. The oscillation (of frequency v) begins to have an effect if T/v is sufficiently large. To be dimensionally correct we should write this as $kT > hv$ (because we must expect T to occur in the combination kT). Here h is a quantity with dimension energy-multiplied-by-time. The rotation (moment of inertia I) becomes significant if $I \cdot T$ is sufficiently large, i.e. if $I \cdot kT > h^2$. Measurements show this h to be of a similar order of magnitude to that introduced through the oscillation. Just as the oscillation contributes to the specific heat, the energy density in a cavity filled with radiation depends on v and T. The energy density is considerable if kT/hv is large enough.

1*b*. While a continuum of states does not yield a natural unit of phase-space (i.e. the (q,p)-plane for one-dimensional systems) and thus leaves us the possibility of introducing an arbitrary factor in the number W of events and an arbitrary additive constant in the entropy $S = k \ln W$, experience shows that the entropy constant has a definite value, which can in fact be determined. It corresponds to the unit h in the (q,p)-plane for a one-dimensional system. The quantum of action h was discovered in the year 1900 for black-body (vacuum) radiation. The links with the specific heats, derived from the oscillations and rotations, were discovered in 1907 and 1913 respectively; gas-entropy in 1913 (Chapter 2).

2. Certain phenomena connected with the absorption of light and X-rays reveal an effect which is particularly noticeable for short-wave (high-frequency) radiation, and which is also irreconcilable with the idea that energy is uniformly distributed over the radiation field. The phenomena we have described were related to quantum theory in 1905 (Chapter 4) by the introduction of the concept of *light quanta* and the recognition that radiation of frequency v could be absorbed and emitted only in amounts of energy hv. This simultaneously explained a whole host of other phenomena that arise when electricity passes through a gas.

3. The *atom* itself gives indications of h. It is quite impossible to give a simple explanation of the chemical properties of the elements: expressed in the periodic system, they have peculiar periodic lengths 2, 8, 8, 18, 18,..., $(2n^2)$. We now know that these numbers are bound up with the spin and PAULI's exclusion principle (Chapter 9) rather than with the simple foundations of quantum theory. But the definite chemical nature of the elements indicates a congruence between the atoms of an element and it also suggests a particular *stability* of atoms. Together with the evidence from gases, this points to a definite *atomic radius*. The question of the stability of an atom and of the radius became a burning one after RUTHERFORD's discovery of the essentially point-like positively charged nucleus with a Coulomb force field (1911). For the constants m and e (electron mass and electronic charge; as the nucleus remains almost at rest, its mass has no effect) cannot be arranged to give a quantity with the dimension of length. For an H-atom with one electron the relationship between centripetal force and Coulomb force

$$m\omega^2 a^3 = e^2$$

(we have put $4\pi\varepsilon_0 = 1$) must be extended by a further assumption which makes no sense in classical terms: that of the angular momentum

$$m\omega a^2 = \hbar$$

(we have assumed a circular path with radius a) so that we can then determine the atomic radius

$$a = \frac{\hbar^2}{me^2}.$$

An equation of this kind was formulated in 1910 (Chapter 4). The explanation of the structure of the atom in terms of quantum theory followed in 1913 (Chapter 5).

4. Line-spectra and the series laws governing them give a wealth of information. The fact that the spectral frequencies are differences between 'terms' of the type

$$v = F(n_1 ...) - F(n_2 ...)$$

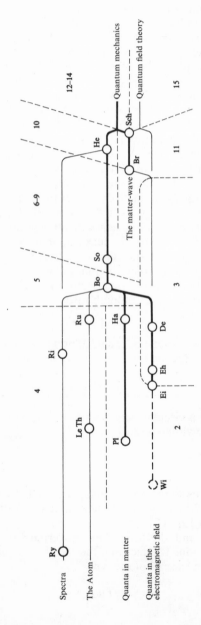

FIGURE 1: PLAN OF THE DEVELOPMENT OF QUANTUM THEORY

(The numbers refer to the Chapters)

gives a very direct indication of h, as does the fact that the complicated systems of frequencies can be expressed as functions of the
simpler of the 'terms'. In 1913 the above equation for v was
recognized to be a special case of the relationship

$$hv = E(n_1 ...) - E(n_2 ...).$$

5. It was assumed in 1923, by analogy with the wave-particle
duality for light, that matter might also possess *wave properties*.
No experimental proof was possible for a number of years
(Chapter 11).

Of all these experimental pointers to h in 1900 only black-body
radiation and spectra were understood, and even those only to a
limited extent. Quantum theory could have developed from either
of these starting points. In fact, it grew from black-body radiation.
Figure 1 gives a rather crude representation of the main lines
along which the subject developed. The chapters that follow (indicated in Figure 1 by numbers and marked off by broken lines)
will show how the lines that represent quanta in matter (PLANCK
and HASENÖHRL), quanta in the electromagnetic field (WIEN,
EINSTEIN, EHRENFEST and DEBYE), the atom (LENARD, THOMSON,
RUTHERFORD) and spectra (RYDBERG, RITZ) were all joined together by BOHR.

A straight path led from BOHR, via SOMMERFELD, to the formulation of quantum mechanics as given by HEISENBERG. Taken
together with the SCHROEDINGER equation, itself a consequence
of DE BROGLIE's wave concept, HEISENBERG's work meant that it
was finally possible to develop a general quantum mechanics.

[1] LORD KELVIN, Phil. Mag. **2**, 1 (1901)

Works on the history of quantum theory
   M. JAMMER, The Conceptual Development of Quantum Mechanics, New
   York 1966
   B. L. VAN DER WAERDEN, Historical Introduction to Sources of Quantum
   Mechanics, Amsterdam 1967, and New York 1968
   A. HERMANN, Introduction to Vols. 1, 3, 4, 5, 7, 10 of Dokumente der Natur-
   wissenschaft, Abteilung Physik, Stuttgart 1962, and Frühgeschichte der Quan-
   tentheorie (1899–1913), Mosbach (Baden) 1969
   R. E. PEIERLS, The Development of Quantum Theory, Contemporary Physics
   6, 129, 192 (1965)
The best source for biographical data is:
   T. S. KUHN, J. L. HEILBRON, P. FORMAN, L. ALLAN, Sources for History of
   Quantum Physics, Amer. Philos. Soc. Philadelphia 1967

2. QUANTUM STATISTICS[1]

Energy Distribution

ONE of the problems dealt with by statistical mechanics is that of the distribution, in thermal equilibrium, of the energy among the various parts of a physical system. This includes the problem of determining the values of $E(T)$, the energy of a body as a function of the temperature. This in turn involves the determination of the specific heat and $w(T,v)dv$, the energy of a radiating black body per unit volume in the frequency interval dv. It also includes the problem of the equilibrium of a gas with its liquid form and of the distribution of gases that react chemically with each other. h appears in all three phenomena. It was recognized in 1900 for black-body radiation, in 1907 for specific heats, and in 1912 for the properties of gases.

Black-Body Radiation

It follows from the concept of thermal equilibrium that the energy density $w(T,v)$ per unit frequency in a radiating black body is independent of the conditions of the walls, of the position, and of the directions of radiation and polarization. The function w of the two variables T and v thus expresses a very general property of nature. In 1893–4, by considering the adiabatic change of frequency through moving mirrors, Wilhelm WIEN saw that this function could be derived from a function of a single variable:

$$w(T,v) = v^3 f\left(\frac{v}{T}\right) \tag{1}$$

(in fact, he wrote the corresponding formula with the wave length λ as the variable). This is 'WIEN's Displacement Law'.[2] If f does not signify simply raising to some power—and that is not possible, thanks to the falling-off at high frequencies indicated by the experiments—it must be possible to write it as a function of a dimensionless variable. As T must arise in the combination kT

this variable would have to be hv/kT, where h has the dimension of energy multiplied by time. As the properties of the radiation field may depend only on the constants c and $4\pi\varepsilon_0$ of the electromagnetic field, and as it is impossible to construct such a quantity h using only these constants, we should have a measurable quantity h that could not be explained in terms of classical physics. Or, to put it another way, the energy density w (or the intensity proportional to it) of the radiation displays a temperature-dependent maximum for a frequency v_m. From (1) it follows that $v_m/T = $ const. or

$$hv_m/kT = 1.$$

At the time this idea was neither expressed nor perhaps even considered. But the radiation constant $b = T\lambda_m$, which was very carefully measured in the years 1897–9 (λ_m differs from c/v_m by a constant scale factor), and which could be written as $T\lambda_m = ch/k$ from dimensional considerations, contains a non-classical quantity h that differs from PLANCK's quantum of action only by a numerical factor.

The function $w(T,v)$ (which could alternatively be written as a function of T and λ) was measured with great care in the last decade of the nineteenth century, in particular at the Physikalisch-Technische Reichsanstalt in Charlottenburg, Berlin. On the basis of these measurements, two formulae for $w(T,v)$ were derived, both of which were to be of great importance in the history of quantum theory. The first was that of WIEN (1896) and the second that of Max PLANCK (October, 1900). These were joined by a third, due to LORD RAYLEIGH (June, 1900), which he derived from theoretical considerations.

Three Radiation Formulae

In 1896 WIEN expressed the measurements in terms of a formula which was equivalent to

$$w(T,v) \sim v^3 \exp(-bv/T). \tag{2}$$

To do this he conceived of a model consisting of molecules that emitted light, and whose velocity v was a function of v. Statistical mechanics gives a factor $\exp(-mv^2/2kT)$ for the frequency (of occurrence) of the various velocities. According to WIEN's displacement law v^2 had to be proportional to v, and w must be of

the form (2).[3] And so, for the very first time, the idea emerged
that radiation was connected with particles of energy $\sim v$. These
were still not exactly the same as EINSTEIN's quanta of light, but
very nearly.

PLANCK also tried to explain black-body radiation. He attempted to derive the second law of thermodynamics from electromagnetic principles. He constructed a model for the emission
and absorption of radiation at the walls of a black body by
assuming the presence of resonators (harmonic oscillators), and
looked for a radiation state that would remain stationary under
emission and absorption. The particular properties of the
material were, of course, of no relevance. In 1899, in a rather
roundabout way, he discovered the formula:

$$w = (8\pi/c^3)v^2\bar{E}(T,v) \tag{3}$$

where $\bar{E}$ was the average energy of the resonators. He no longer
calculated $\bar{E}$ from the equipartition theorem, as this would have
given $\bar{E} = kT$ and $w \sim v^2 T$, contrary to the experimental evidence. He made no statistical assumption about $\bar{E}$. Instead, he
sought an expression for the entropy of an oscillator with
frequency v. Using the 'definition'

$$S = -\frac{E}{bv}\ln\frac{E}{ea\cdot v} \tag{4}$$

he was able to demonstrate the stationary nature of this radiation
state and obtain WIEN's formula (2).[4] His derivation of the formula was equivalent to the following argument (using the average
value $\bar{E}$):

$$\frac{1}{T} = \frac{dS}{d\bar{E}} = -\left(\frac{1}{bv}\ln\frac{\bar{E}}{eav} - \frac{1}{bv}\right) = -\frac{1}{bv}\ln\frac{\bar{E}}{av} \qquad \bar{E} = ave^{-\frac{bv}{T}}.$$

It also followed that:

$$\frac{d^2 S}{dE^2} = -\frac{1}{bvE}.$$

It was very important that this second derivative should be
negative, so that there was an increase in entropy as a result of
the interchange of energy between two oscillators. Wrongly,
PLANCK assumed that the stationary property was purely a result
of his assumption (4). In fact, any expression with a negative
second derivative would have given the same result. And so he

was satisfied that WIEN's radiation formula had a sound theoretical basis.[5]

In 1900 RAYLEIGH investigated the distribution of energy among the characteristic oscillations of the electromagnetic field in a black body.[6] Although the characteristic oscillations are uniformly distributed over the v spectrum in the case of a vibrating string, and in a two-dimensional linearly oscillating system the number $Z(v) \, dv$ in the interval dv is proportional to v, for a three-dimensional black body we get:

$$Z(v) \sim v^2 .$$

It was not until 1905 that RAYLEIGH published the factor

$$Z(v) = \frac{8\pi}{c^3} v^2 .$$

Thus he derived the expression

$$w(T,v) \sim v^2 \, \bar{E}(T,v)$$

for the energy density w, where $\bar{E}$ now denoted the average energy of a black-body oscillation. (With a factor $8\pi/c^3$ this would have been precisely PLANCK's formula (3), though here for electromagnetic oscillators.) He then took the equipartition theorem for $\bar{E}$ and so derived the radiation formula

$$w(T,v) \sim v^2 \, k \, T. \tag{5}$$

With a factor $8\pi/c^3$ this would have read

$$w(T,v) = \frac{8\pi}{c^3} v^2 \, k \, T. \tag{6}$$

Of course, formula (5) contradicted the experimental results, but RAYLEIGH thought that it might possibly hold for long waves (small v).

More precise measurements of black-body radiation raised doubts as to the strict validity of WIEN's formula (2). In 1900 Heinrich RUBENS and Ferdinand KURLBAUM at the Physikalisch-Technische Reichsanstalt discovered significant deviations at low frequencies. PLANCK, who heard of these results, was therefore obliged to look for a different expression for the entropy of his resonators. In October 1900 he modified[7] his earlier formula

$$\frac{d^2 S}{d\bar{E}^2} \sim -\frac{1}{\bar{E}}$$

by putting

$$\frac{d^2 S}{d\bar{E}^2} = -\frac{\alpha}{\bar{E}(\beta+\bar{E})}. \tag{7}$$

He integrated this to obtain

$$\frac{1}{T} = \frac{dS}{d\bar{E}} = \frac{\alpha}{\beta}\ln\frac{\beta+\bar{E}}{\bar{E}}$$

from which it followed that

$$\bar{E} = \frac{\beta}{e^{\frac{\beta}{\alpha T}}-1} \qquad w = \frac{8\pi}{c^3}\frac{v^2\beta}{e^{\frac{\beta}{\alpha T}}-1}.$$

Because of WIEN's displacement law, $\beta \sim v$ and α must be independent of v. PLANCK still wrote his result in terms of λ, but it was equivalent to

$$w = \frac{8\pi}{c^3}\frac{hv^3}{e^{\frac{hv}{kT}}-1}. \tag{8}$$

PLANCK's expression (7) was an interpolation between $\sim -1/\bar{E}$ and $\sim -1/\bar{E}^2$ for the value of $d^2S/d\bar{E}^2$. The choice of the latter would have led to $w \sim v^2 T$, RAYLEIGH's formula. Thus PLANCK made a simple interpolation between the WIEN and RAYLEIGH radiation formulae. His formula was in very good accord with the measurements.

Thus, in October, 1900, there were three radiation formulae: RAYLEIGH's, which stated that $w \sim v^2 kT$, theoretically sound and correct for small values of v, WIEN's, $w \sim v^3 \exp(-hv/kT)$ (rewritten in terms of v), not yet understood theoretically, and true for large values of v, and finally that of PLANCK, $w \sim v^3/[\exp(hv/kT)-1]$, which was also not understood theoretically, but which was true for all values of v. All three satisfied WIEN's displacement law $w = v^3 f(v/T)$.

The Birth of Quantum Theory

Between 19th October and 14th December 1900 PLANCK was working on a theoretical explanation of his interpolated radiation formula. He knew the relationship

$$w(T,v) = \frac{8\pi}{c^3}v^2\bar{E}(T,v) \tag{3}$$

for the average energy $\bar{E}$ of the oscillators in his model. RAYLEIGH had discovered the same relationship in a simpler way for the average energy of the characteristic waves of the electromagnetic field. PLANCK now approached the question of the average energy of the oscillators as that of the statistical distribution of the energy among the oscillators of a particular frequency v.[8] He distributed P energy quanta of size ε among N oscillators, so that the latter could have the energy $P_1\varepsilon$, $P_2\varepsilon$,..., where P_1, P_2,... are natural numbers. He regarded an event as consisting of a particular sequence of occupation numbers P_1, P_2,... of the oscillators. In the terminology later adopted, PLANCK's calculation amounted to the following: the events that arise are the quantum states $E_l = P_l\varepsilon$ taken by the individual oscillators. PLANCK was now in a position to calculate the average energy $\bar{E}$ of the oscillators in thermal equilibrium from their entropy, and obtained

$$\bar{E} = \frac{\varepsilon}{e^{\frac{\varepsilon}{kT}} - 1}.$$

In order to preserve agreement with WIEN's displacement law, he had to assume $\varepsilon \sim v$. He put

$$\varepsilon = hv. \tag{9}$$

There are two points of great significance in PLANCK's work, which was to form the foundation of quantum theory. PLANCK chose a quantum of energy ε of finite magnitude. If he had let ε tend to zero, he would have obtained $\bar{E} = kT$, the equipartition theorem, and RAYLEIGH's formula. Moreover, PLANCK's method of counting was different from what would have been given by BOLTZMANN's theory. According to BOLTZMANN, a state would be defined by declaring in which box the first quantum lay, in which the second one lay, and so forth. This would lead to WIEN's formula. The underlying assumptions of PLANCK's calculations were discovered by PLANCK himself in 1906 and by L. NATANSON in 1911. PLANCK's method corresponds to what we now call BOSE particle statistics. Of the two noteworthy points in PLANCK's work, the finite nature of the quanta leads to the deviation from the classical RAYLEIGH formula, and the non-BOLTZMANN method of counting events leads to the deviation from WIEN's formula.

By comparing the measured energy densities (or radiation intensities) with PLANCK's formula it was possible to calculate h

and h/k. PLANCK was immediately able to determine h/k and, using $\int w(T,v)\,dv$, he obtained another combination of h and k. He was thus able to obtain the most accurate value yet found for the constant k. From this he calculated AVOGADRO's number and the value of the elementary charge, and also that of the new universal constant h.

Thus at the outset of quantum theory we encounter three important figures: Wilhelm WIEN, LORD RAYLEIGH and Max PLANCK. WIEN's displacement law and his radiation formula already implicitly contain the quantity h. PLANCK did not at first recognize the radical significance of these laws. Instead, he attempted to place WIEN's formulae on a firm theoretical foundation on the basis of contemporary physical theory. In doing so he overlooked the equipartition theorem. RAYLEIGH showed that this theorem led to an impossible result. On the other hand, PLANCK saw the central significance of entropy and so derived his radiation formula. The only way of explaining it led him to introduce the energy quantum hv.

Repercussions

The first perceptible repercussions of PLANCK's theory occurred in the summer of 1905. They came in the form of brief notes by RAYLEIGH and Sir James JEANS.[9] JEANS had attempted to solve the problem of specific heats by assuming a very slow interchange of energy between specified degrees of freedom. RAYLEIGH was aware of a similar difficulty in the case of black-body radiation and guessed that it would be possible to understand specific heats once black-body radiation had been explained. He did not accept JEANS's way out of the difficulty. RAYLEIGH next wrote his radiation formula with the factor $8\pi/c^3$ and posed the question of how it was possible for PLANCK to obtain a result other than $w \sim v^2 T$ in the framework of BOLTZMANN's theory. JEANS accused PLANCK of using incorrect statistics—he should really have let h tend to zero. For JEANS the properties of radiation signified that the aether was not in equilibrium. For RAYLEIGH they indicated that the equipartition theorem was not valid even for thermal equilibrium.

PLANCK's hypothesis consisted in a distribution of energy quanta among real, concrete oscillators. A slight change of emphasis led to the assertion that harmonic oscillators had discrete energies $E = hvn$. Such oscillators might well be real ones, but they could equally well be electromagnetic waves. Moreover, there was also the WIEN particle model for the interpretation of black-body radiation, the energy of the particles being a function of the radiation frequency. For each of these different points of view there was a different course along which quantum mechanics evolved: energy quanta, light quanta, the energy states of material oscillators, and those of electromagnetic oscillators. We shall trace these in that order. For the time being, everything remains within the realm of statistics, and the harmonic oscillator is the only problem we shall deal with.

The Statistics of Indistinguishable Quanta

In PLANCK's statistics the events were the numbers of quanta ε (all treated equally) in the individual oscillators at any one frequency. This was rather different from the distribution of independent quanta ε among the oscillators, whose events would be the oscillators occupied by the individual quanta. PLANCK's method could be understood by taking the energy states $E = hvn$ of an oscillator as the events. This was the approach later taken by PLANCK. It was, however, also possible to interpret it in the way NATANSON did in 1911, as a distribution of indistinguishable quanta among the distinguishable but identical boxes of finite size. No new events arise if we permute the quanta in any one box.[10] For example, take $P = 3$ identical quanta, and $N = 2$ identical boxes. Independent, distinguishable quanta give us $2^3 = 8$ events. Independent and indistinguishable quanta give

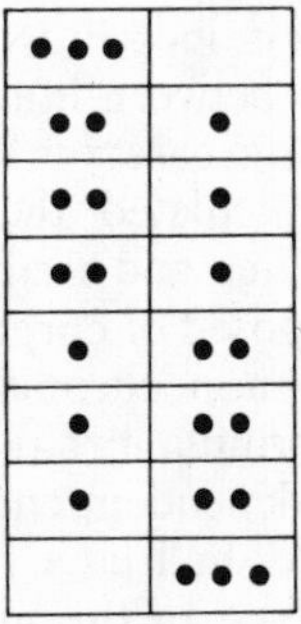

FIGURE 2: DISTRIBUTION OF THREE PARTICLES BETWEEN TWO BOXES

rise to only four events (Figure 2). In general, distinguishable quanta give N^P events, while indistinguishable ones give

$$W = \frac{(N+P-1)!}{P!(N-1)!}$$

events. This last formula was used by PLANCK. It is easiest to understand this formula by illustrating each event in the following way, for example by showing 2, 1, 4, 0, 3 quanta in 5 boxes, as

$$\cdot\cdot\,|\,\cdot\,|\,\cdot\cdot\cdot\cdot\,|\quad|\,\cdot\cdot\cdot$$

(this was suggested by EHRENFEST). There are $(N + P - 1)!$ permutations of these $N + P - 1$ elements; no other arrangement is given by the $P!$ permutations of the points and the $(N - 1)!$ permutations of these lines. This method of enumeration, classified by NATANSON, is precisely that later used by BOSE for light quanta, and now called BOSE statistics. In the transition to arbitrarily small boxes (i.e., to classical statistics) the number of events in a distribution tends towards that given by BOLTZMANN statistics, corrected by a factor $P!$ in the denominator, and a factor in the numerator that depends on the arbitrary size of the boxes.

Light Quanta

WIEN's model with particles of energy $\varepsilon \sim \nu$ was, in 1905, developed by Albert EINSTEIN in a much more precise form as the hypothesis of energy quanta $\varepsilon = h\nu$ in radiation.[11] He called this an 'heuristic point of view' and thus hoped to use it to gain new insight. He briefly examined PLANCK's radiation theory and proved the contradiction between it and established theory. He then analysed the limiting case of WIEN's radiation formula $w \sim \nu^3 \exp(-h\nu/kT)$, calculated the corresponding entropy, just as WIEN had done earlier, and established that it behaved like the entropy of a gas composed of corpuscles with energy $\varepsilon = h\nu$. His result ran as follows: monochromatic radiation in the region of validity of WIEN's formula behaves thermodynamically as though it consisted of independent energy quanta of magnitude $h\nu$ (he wrote $R\beta\nu/N$). We shall later (in Chapter 3) discuss the application, due to EINSTEIN, of this model to luminescence, the photoelectric effect, and to ionization.

One year later EINSTEIN attempted to explain PLANCK's radiation formula.[12] He did not yet recognize the corresponding
entropy as being caused by particles and waves (cf. Chapter 3).
He was, however, able to explain it so long as he did not assume
a continuum of possible energy values, but only the discrete
values $E = h\nu n$. PLANCK's formula $w \sim \nu^2 \bar{E}$ was, however, no
longer fundamental. 'PLANCK appears to have introduced a new,
hypothetical element into physics—the hypothesis of light
quanta.' And that was how EINSTEIN reconciled himself to
PLANCK's radiation theory.

The Energies E(n) of an harmonic Oscillator

The distribution of energy quanta $\varepsilon = h\nu$ among oscillators is
equivalent to the assumption that oscillators may assume only
the energies

$$E = h\nu n. \tag{10}$$

In PLANCK's 1906 book about the theory of heat radiation this
assumption and the enumeration of these energy levels as 'events'
in statistical thermodynamics receive a new interpretation—at
first it is a kind of *obiter dictum*.[13] In the (x,p) phase plane the
point representing the motion of the system traces out an ellipse
with equation:

$$\frac{p^2}{2m} + \frac{m\omega^2 x^2}{2} = E$$

and area—'phase extension'—$\Phi = \pi ab$, where the semi-axes a
and b are given by

$$\frac{m\omega^2 a^2}{2} = \frac{b^2}{2m} = E.$$

We thus have

$$\Phi = 2\pi \frac{E}{\omega} = \frac{E}{\nu}.$$

PLANCK's formula accordingly gives

$$\Phi = hn. \tag{11}$$

Phase regions of size h are gathered into a single event. We note
that if h were small compared with the phase, the number of

events would be almost a continuum, and the number of events would become proportional to the phase. We see classical, BOLTZMANN statistics as the limiting case of PLANCK's theory.

PLANCK's book led Paul EHRENFEST to make an observation. He had already, in 1905, recognized the important elements in PLANCK's theory as being (i) the fact that an expression for the entropy S which ensured the stationary nature of the radiation state need not be the only one, (ii) the fact that PLANCK had assumed a finite quantum ε, and (iii) that the events were enumerated in a particular way.[14] And now, in 1906, he showed that boundary conditions could be introduced in phase space, that these would influence the radiation density, and that a theory of radiation was not satisfactory until one of the conditions was made relevant and meaningful. We can derive PLANCK's radiation formula by allowing only the values $E = hvn$ of the energy and counting them all as of equal status.[15] EHRENFEST thus treated the electromagnetic waves in a black body as oscillators.

In 1907 EINSTEIN calculated the average energy of an harmonic oscillator with energy levels $E_n = hvn$ by direct summation[16] (see the Appendix).

$$\bar{E} = \frac{\Sigma E_n e^{-\beta E_n}}{\Sigma e^{-\beta E_n}} = \frac{hv}{e^{\beta hv} - 1} \qquad \beta = \frac{1}{kT}. \tag{12}$$

This calculation of the average values was certainly facilitated by the publication in 1902 of J. Willard GIBBS's book on the foundations of statistical mechanics. A German translation appeared in 1905. EINSTEIN's calculation is published in the same paper as that in which he applied PLANCK's hypothesis to the specific heat of a rigid body. He idealized a crystal consisting of N identical atoms by a system of $3N$ harmonic oscillators of the same frequency v, and thus obtained

$$E = 3N \frac{hv}{e^{\frac{hv}{kT}} - 1} \tag{13}$$

for the energy of the oscillators of a crystal. The specific heat calculated from this displayed the DULONG-PETIT law as a limiting case for high temperatures; the higher the value of v, the greater the temperature. In this way EINSTEIN was able to explain the well-known fact that the elements B, C, Al, Si had specific heats essentially lower than those given by the law. EINSTEIN thus effectively accepted PLANCK's quantum theory in

1907. Hardly anything was then known of the decline of specific heats at very low temperatures. Around 1910, however, specific heats were measured at low temperatures. They were not in particularly good agreement with EINSTEIN's formula. Bearing in mind that an oscillating crystal comprises oscillators of various frequencies DEBYE in 1912 was able to derive a theory which accorded well with the experimental evidence.[17] He assumed the distribution of frequencies to be the same as for an oscillating continuum. He restricted their number to that attainable in crystal lattices. In this way he produced a simple and elegant theory. Almost simultaneously Max BORN and Theodor von KARMAN used lattice theory for a similar calculation, more detailed but consequently all the more accurate.[18]

Perhaps the neatest derivation of PLANCK's radiation formula was given by DEBYE in 1910.[19] He used RAYLEIGH's formula

$$Z(v) = \frac{8\pi}{c^3} v^2$$

for the number of electromagnetic waves per unit volume in a black body. In the formula

$$w(T,v) = Z(v)\bar{E}(T,v)$$

he worked out the average value $\bar{E}$ once, classically, equal to kT, that is, to RAYLEIGH's assumption, and once, using quantum theory, from the energy levels $E = hvn$ which, according to (12) gives PLANCK's formula.

EHRENFEST was also happy to use the idea of electromagnetic oscillations of a black body. In 1911 he put into practice his observation of 1906 and tried to discover precisely which features of the light quantum hypothesis were really fundamental to the theory of radiation.[20] In doing this he made use of a weight function $G(E/v)$ for the enumeration of states. The decrease in energy density for large values of v signified that the energy value $E = 0$ had a special weighting of an order of magnitude normally associated with a finite interval. A decrease $\sim \exp(-hv/kT)$ indicated that there was no weight in the neighbourhood of $E = 0$. EHRENFEST went on to emphasize that the acceptance of equal probabilities for the energy values hvn led to the PLANCK formula, but not the assumption that the energy quanta $\varepsilon = hv$ were distributed with equal probability among the different oscillators. This occurred at approximately the same time as the observation by NATANSON mentioned above.

h *as a Unit of Phase Extension for Periodic Motion*

How was it possible to get from the energy levels $E = hvn$ of an harmonic oscillator to the levels $E(n)$ of other systems, in the first instance to one-dimensional ones? PLANCK had so far applied his 1906 method (of treating whole regions of the x,p-plane, the phase-plane, of area h, as a single event) only to the harmonic oscillator. But his method could be generalized. Steps towards such a generalization emerged from the discussions at the Solvay Congress in Brussels in 1911.[21]

This was the first of a series of meetings of physicists that took place usually every three years. Its subject was 'The Theory of Radiation and Quanta'. Among those present were Hendrik Antoon LORENTZ, PLANCK, WIEN, Walther H. NERNST, Arnold SOMMERFELD, Ernest RUTHERFORD, Fritz HASENÖHRL, JEANS, EINSTEIN; EHRENFEST was not there, nor were Peter DEBYE and Niels BOHR, who were still quite young men. LORENTZ's opening speech emphasized the failure of classical mechanics: the physics of the future would be very different. The old theory led to the equipartition of energy among the various degrees of freedom for radiation and for specific heat. 'We do not understand why a lump of iron does not glow at room temperature.' Even the idea that certain degrees of freedom attain equilibrium only after a very long time was unsatisfactory. 'None of which is exactly helpful.'

PLANCK now regarded phase-extension as the crucial point. The old kind of statistics that gave the number of events as proportional to the phase area $dp \, dx$ led to a radiation energy $\sim v^2 T$. The quantum hypothesis, on the other hand, counted a region

$$\iint dx \, dp = h$$

as a single event. The validity of quantum theory was clearly irreconcilable with that of classical mechanics. The distinction between physics and chemistry was seen thus: 'physics' obeyed classical dynamics, being valid for whole molecules and free electrons, and it explained 'physical forces'. The structure of atoms and molecules belonged to 'chemistry', and 'chemical forces' belonged to quantum theory. The quantum-theoretical interpretation of chemical forces was a programme which was to become possible in 1926.

The discussion at the Solvay Congress came close to the idea that

$$\Delta \oint p\,\mathrm{d}x = h$$

for all periodic motion with one degree of freedom. By '$\oint$' we mean an integral over one whole period. But we detect a certain inhibition when we read the Congress report. They were unable to escape from the relationship

$$E = h\nu n.$$

LORENTZ attempted to deal with the rotator by means of the formula

$$E(n) = h\nu(n)n$$

as did BJERRUM in 1912 and EINSTEIN and STERN in 1913.[22] HASENÖHRL, however, solved NERNST's problem of the quantum states of a rotator as follows: let us assume that the frequency of a rotator depends on its energy, distinguishing this case from that of the harmonic oscillator. Equal phase areas then gave unequal energy steps and equal energy steps gave unequal phase areas. The first of these possibilities would certainly appear to correspond to PLANCK's ideas. Actually, if we read the Congress report, we now expect to find the theory of a rotating body with a fixed axis

$$\Phi = 2\pi P = hn \qquad E = \frac{P^2}{2I} = \frac{h^2 n^2}{8\pi^2 I}$$

(this was first written by EHRENFEST in 1913). Instead of this, a few pages later we discover the theory of a rotating body with a free axis, a system with several degrees of freedom, for which, once again, $\Phi \sim E$ and for which HASENÖHRL gives $E \sim n$.

But HASENÖHRL had the right idea—that $\Delta\Phi = h$. At the Karlsruhe meeting (also in 1911) he considered the relationship between $\nu(E)$ and $E(n)$.[23] For the three uniquely related quantities. Φ, E, ν for one degree of freedom, the relationship

$$\mathrm{d}E = \nu\,\mathrm{d}\Phi \qquad (14)$$

holds. (He probably learned that from BOLTZMANN.) Thus the phase plane is divided by curves $E_n(x,p) = \text{const.}$ into regions such that

$$\int_{E_0}^{E_1} \frac{\mathrm{d}E}{\nu(E)} = \int_{E_1}^{E_2} \frac{\mathrm{d}E}{\nu(E)} = \cdots = h. \qquad (15)$$

The first example considered by HASENÖHRL was that of an anharmonic oscillator with a linear $v(E)$; the frequencies of the quantum states are then likewise linear in n:

$$v(n) = v_0(1 + \alpha n).$$

DESLANDRES had found similar series of frequencies in band spectra. The second example is interesting for the mistake that HASENÖHRL made. He attempted to deduce the energy levels $E(n)$ of a hydrogen atom from the Balmer formula for the frequencies:

$$v = \frac{n^2 - 4}{4n^2} Rc.$$

By integrating

$$v = \frac{dE}{d\Phi} = \frac{dE}{h\,dn} = \frac{n^2 - 4}{4n^2} Rc$$

he obtained

$$E = \frac{n^2 + 4}{4n} Rch$$

a formula which, as we know, has no connection whatsoever with reality. HASENÖHRL did not spot the difference

$$v = \left(\frac{1}{4} - \frac{1}{n^2} \right) Rc$$

in the Balmer formula although Carl RUNGE had given it a similar form to this in 1888, and RYDBERG had in 1890 written the alkali spectra series in the form

$$v = A - \frac{B}{(n+\alpha)^2}.$$

Indeed, Walter RITZ had given a general statement of the combination principle

$$v = F(n_1 \ldots) - F(n_2 \ldots)$$

in 1908, and Friedrich PASCHEN had discovered the hydrogen series

$$v = \left(\frac{1}{9} - \frac{1}{n^2} \right) Rc$$

in 1909. If HASENÖHRL had taken serious account of the difference, he would surely have concluded that

$$v = \frac{\Delta E}{\Delta \Phi} = \frac{\Delta E}{h} \qquad E = -\frac{Rch}{n^2}.$$

EINSTEIN's formula

$$hv = \Delta E$$

had after all been known since 1905.

Like HASENÖHRL in 1911, DEBYE in 1913 also took as fundamental to quantum theory the fact that a region of phase space

$$\Delta \Phi = \Delta \oint p \, dx = h$$

was replaced by a single event in the statistical enumeration.[24] For the equation of state of a rigid body he needed to use the average displacement of an asymmetrical anharmonic oscillator. Whereas this is $\sim kT$ according to classical statistics, quantum theory gives the factor $hv/[\exp(hv/kT) - 1]$.

In the same year Paul EHRENFEST gave the values

$$E = h^2 n^2 / 8\pi^2 I$$

for the energy levels of a rotating body, which therefore means that

$$\Phi = 2\pi P = \pm hn \tag{16}$$

i.e., phase regions of size h.[25] In a footnote he suggested that the phase extension Φ of a motion did not change if the oscillations were transformed into orbits by the gradual weakening of an external field (in the phase plane this would mean the transformation of ellipses into lines). In his detailed paper of 1913 on this adiabatic invariance of

$$\Phi = 2\oint E_{kin} \, dt = 2\frac{\overline{E_{kin}}}{v}$$

(E_{kin} is the kinetic energy, $2\oint E_{kin} \, dt$ is the same as $\oint p \, dx$) he assumed, more generally, that $\Phi = hn$ 'as this is true for the oscillator'. Anyway, he now changed (16) for a rotating body to $4\pi P = \pm hn$; this is bound up with the fact that the relationship between oscillation and rotation is not uniquely determined if the oscillation becomes an orbiting motion.

The quantum prescription for periodic motion with one degree of freedom

$$\Phi = \oint p\,\mathrm{d}x = hn$$

seemed to be well established in 1913.

h *as a Unit in Translation*

The collapse of a region of phase space of size h into a single event was characteristic of the quantum statistics of periodic systems with one degree of freedom. It was not long before a similar role was found for h in pure translation.

The thermodynamic definition of entropy $\mathrm{d}S = \mathrm{d}Q/T$ leaves a free additive constant in S. It follows for an ideal gas, per mole, that

$$S = \int \frac{mc_p\,\mathrm{d}T}{T} - R\ln p + \text{const} \tag{17}$$

(where mc_p is the molar specific heat at constant pressure p, and R is the gas constant) and for a monatomic gas that

$$S = R\ln \frac{T^{5/2}}{p} + \text{const.} \tag{18}$$

It is possible to use $\Delta Q/T$ to measure the change in entropy for fusion, for vaporization and for chemical changes. According to the NERNST heat theorem (1905) all entropy differences tend to zero for condensed systems, and the specific heats thus tend to zero with the temperature. If the measurements on the condensate extend to sufficiently low temperatures it is possible to extrapolate to zero temperature, and using the specific heats, the latent heats of fusion and of vaporization it is possible to determine the entropy $S - S_0$, where S_0 denotes the entropy of the condensate at $T = 0$. PLANCK in 1910 put this constant S_0 equal to zero, and thus gave the NERNST heat theorem the following extended form: the entropy of a condensed system is equal to zero at a temperature of absolute zero.[27] It is then possible to determine the values of the constants in (17) and (18) by experiment and to calculate the equilibrium states of mixtures of reactive gases. The equilibria calculated in this way agreed with those measured, and any remaining inconsistencies were cleared up by more precise measurements in due course.

If we calculate the entropy of a system from the formula $S = k \ln W$, where W is the phase extension, the arbitrary unit of phase extension (which we may take as small as we please) leads to an arbitrary phase constant. For a gas of N independent particles the transition from a box size Ω to one of ω causes the number W of events to be multiplied by $(\Omega/\omega)^N$. It thus increases the entropy S by $kN \ln (\Omega/\omega)$. For a monatomic gas the entropy per mole for a box size ω is

$$S = R \ln \frac{(kT)^{5/2}(2\pi\mu)^{3/2}}{p\omega}$$

(μ is the molecular mass). By comparing the empirical values of the entropy constants with the predicted ones (for example in the case of mercury vapour) we can determine the value of what must be a natural constant. Otto SACKUR thought of this in 1911, and in 1912 he and TETRODE discovered that for every degree of freedom h was the natural unit and S_0 was equal to zero.[28] At roughly the same time PLANCK realized that the deeper significance of the NERNST heat theorem lay in the fact that at absolute zero $S = 0$ and $W = 1$, so that the state of a system was completely determined. He had not, however, grasped the idea of h as a unit of phase space for translation.[29]

To sum up, by about 1913 the development of quantum theory had led to a kind of quantum statistics. The general principles of statistical physics—the definition of entropy $S = k \ln W$, even for non-equilibrium states, and the definition of temperature as $\exp(-E/kT)$ for equilibrium—both remained untouched. On the other hand the method of counting events, the determination of W, was quite different in quantum statistics.

There is a natural unit for phase extension: h *for each degree of freedom. For periodic motion with one degree of freedom it leads to a restriction to the states* $\Phi = \oint p \, dx = hn$.

[1] M. PLANCK, Wissenschaftliche Selbstbiographie, Leipzig 1948
L. ROSENFELD, La première phase de l'évolution de la théorie des quanta, Osiris **2**, 149 (1936)
M. J. KLEIN, M. Planck and the beginning of the quantum theory, Arch. Hist. Ex. Sci. **1**, 459 (1962)
[2] W. WIEN, Sitz. Ber. Berlin **1893**, 55, Wied. Ann. **52**, 132 (1894)
[3] W. WIEN, Wied. Ann. **58**, 662 (1896)
[4] M. PLANCK, Sitz. Ber. Berlin **1899**, 440, Ann. d. Phys. **1**, 69 (1900)
[5] M. PLANCK, Ann. d. Phys. **1**, 719 (1900)
[6] LORD RAYLEIGH, Phil. Mag. **49**, 539 (1900)

[7] M. PLANCK, Verh. D. Phys. Ges. **2**, 202 (1900)

[8] M. PLANCK, Verh. D. Phys. Ges. **2**, 237 (1900)

[9] LORD RAYLEIGH, Nature **71**, 559, 72, 54, 243 (1905)
J. H. JEANS, Nature **71**, 607, **72**, 101, 293 (1905), Proc. Roy. Soc. A **76**, 545 (1905)

[10] L. NATANSON, Phys. Z. **12**, 659 (1911)

[11] A. EINSTEIN, Ann. d. Phys. **17**, 132 (1905)

[12] A. EINSTEIN, Ann. d. Phys. **20**, 199 (1906)

[13] M. PLANCK, Vorlesungen über die Theorie der Wärmestrahlung, Leipzig 1906

[14] P. EHRENFEST, Wien. Ber. m.-n. Kl. **114**, 1301 (1905)

[15] P. EHRENFEST, Phys. Z. **7**, 528 (1906)

[16] A. EINSTEIN, Ann. d. Phys. **22**, 150 (1907)

[17] P. DEBYE, Ann. d. Phys. **39**, 789 (1912)

[18] M. BORN and TH. V. KÁRMÁN, Phys. Z. **13**, 297 (1912), **14**, 15 (1913)

[19] P. DEBYE, Ann. d. Phys. **33**, 1427 (1910)

[20] P. EHRENFEST, Ann. d. Phys. **36**, 91 (1911)

[21] La théorie du rayonnement et les quanta. Réunion à Bruxelles 1911, Paris 1912

[22] N. BJERRUM, Nernst-Festschrift p. 90, Halle 1912
A. EINSTEIN and O. STERN, Ann. d. Phys. **40**, 551 (1913)

[23] F. HASENÖHRL, Phys. Z. **12**, 931 (1911)

[24] P. DEBYE in: Vorträge über die kinetische Theorie der Materie und Elektrizität, Leipzig 1914

[25] P. EHRENFEST, Verh. D. Phys. Ges. **15**, 451 (1913)

[26] P. EHRENFEST, Proc. Amst. **16**, 591 (1913)

[27] M. PLANCK, Vorlesungen über Thermodynamik 3. A., Vorwort, Leipzig 1910

[28] O. SACKUR, Ann. d. Phys. **36**, 958 (1911), Verh. D. Phys. Ges. **14**, 951 (1912)
H. TETRODE, Ann. d. Phys. **38**, 434, **39**, 255 (1912)

[29] M. PLANCK, Wärmestrahlung, 2. A. (p. 131), Leipzig 1913

3. LIGHT QUANTA[1]

Beyond Quantum Statistics

THE quantum hypothesis was first applied to temperature-dependent properties. It explained the failure of equipartition and helped to clarify the significance of the entropy constant. So it was initially part of statistical physics. But whereas classical statistics had measured the number of events by the phase extension $dx\, dp$ (for one degree of freedom), in *'quantum statistics'* a region of phase space

$$\Delta\Phi = \Delta \oint p\, dx = h \tag{1}$$

was counted as a single event. That is roughly how DEBYE put it in 1913. In the limit, as $h \to 0$, quantum statistics tended to classical statistics. From (1) we obtain

$$\Phi = E/v = hn$$

for the harmonic oscillator, and

$$\Phi = 2\pi P = hn$$

for the free rotator.

The variable v was still not used in its dynamic sense. For the oscillator it was at first merely a characteristic of some property of the system accurately measurable in the case of a black body but only roughly for a rigid body. For the rotator it was the energy levels that were of interest, particularly as the expression for them was independent of v. Quantum theory had not yet evolved into quantum dynamics.

One of the steps made by early quantum theory led beyond mere statistics. This was EINSTEIN's 1905 hypothesis of light quanta. Certainly it arose from within quantum statistics, but it was immediately applied to physical phenomena, actually to those involved in the interaction of light and matter. Some of these are rightly regarded as the simplest of all quantum phenomena, e.g. the photoelectric effect, the release of electrons by the action of light, and the *'Bremsstrahlung'*, the production of X-rays by the deceleration of electrons. Moreover, the light quantum hypothesis was a very fruitful concept, even if it was taken seriously only by a handful of people. Within this framework the problem of understanding quantum theory became the problem of 'duality'—the utterly implausible dual nature of light. How were the concepts of waves and particles to be reconciled? Not until 1927 was this question answered for both light and matter.

The simplest forms of matter involved in the interaction of light and matter are electrons and atoms. Thus the energies with which we have to deal in the case of light are $h\nu$. For free electrons they are $\frac{1}{2}mv^2$ and for atoms they are the energy levels $E(n)$. We can take the diagram given below as our basis.

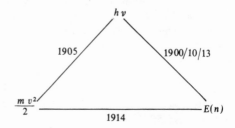

The interaction of light and electrons became part of quantum theory in 1905. These quantum phenomena were most striking in the case of X-rays. The interaction of electrons and atoms, electron collision, was convincingly treated as a quantum phenomenon in 1914. The essential conditions for progress were (a) the ability to experiment with cathode rays and (b) a knowledge of X-rays—matters that were at that time treated under the general heading of 'the conduction of electricity by gases'.

Cathode Rays

Cathode rays had been discovered in 1858 soon after the technological prerequisites had been satisfied. It was quickly established that they consisted of negatively charged particles. The exact determination of the ratio e/m was achieved in 1897 using electric and magnetic fields. Assuming that the particles carried an elementary charge, which could be calculated by electrolysis, knowing the number of molecules in a mole, the very small value of the mass was determined. The particles were called electrons. The frequencies of radiation in the visible and ultraviolet region, which one had to ascribe to motion within the atoms, also suggested by their magnitude that particles of very low mass were involved. In 1897, LORENTZ explained the frequency splitting (ZEEMAN effect) that occurs in an external magnetic field as an additional motion of particles of low mass, caused by the magnetic field. He discovered the same ratio e/m as held for cathode rays. Thus electrons were recognized as the building blocks of atoms, and it was accepted that spectral frequencies must be connected in some way with their motion. Around 1900 it was possible to carry out quite accurate experiments using cathode rays. Philipp LENARD was particularly successful. He caused very fast cathode rays to pass through matter and discovered that atoms must be almost entirely empty. In 1900 he was able to explain a phenomenon discovered in 1888 by Wilhelm HALLWACHS by proving that light of sufficiently high frequency released electrons from the surfaces of metals. The detailed investigation of this phenomenon led him, in 1902, to a very striking result: the velocity of the electrons released did not depend on the intensity of the impinging light, but only on its frequency. Even at low density, electrons were released with an energy that was often much greater than that of the light absorbed by their atomic surroundings. He therefore assumed that the energy of the released electrons was not, or was not entirely, the result of the radiation, but that light was somehow able to release energy that was already present. It was possible for EINSTEIN, in 1905, to explain this phenomenon, in terms of his light quantum hypothesis, as a direct conversion of the energy from light into matter according to the equation

$$h\nu = \tfrac{1}{2}mv^2 + P \tag{2}$$

where P is the work that has to be done in order to release the electrons.

It was soon recognized that 'β-radiation' of radioactive elements also consisted of electrons.

Electric Discharge in Gases. The Collision of Electrons

Around the turn of the century attempts were made to understand what happened when electricity was passed through rarefied gases. Figure 3 shows the typical luminosity between the anode and cathode of a discharge tube as then in common use. Below

(Figure 3) there is a graph of the strength of the electric field plotted along the tube. There were good grounds for the assumption that the luminosity was connected with the ionization of atoms or molecules by the collision of electrons accelerated by the electric field (for example, J. J. THOMSON 1903). In 1903 Johannes STARK had guessed that there was a threshold of kinetic energy of colliding electrons below which they could not ionize an atom. In 1902 LENARD attempted to measure ionization energies. He found the value 11 eV for several gases, but was not completely able to dispose of mercury (with ionization energy 10·4 eV).

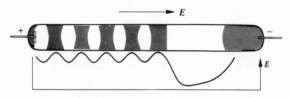

FIGURE 3:   TYPICAL DISCHARGE

E. GEHRCKE and Rudolf SEELIGER discovered in 1912 that there was a qualitative connection between the velocity of the colliding electrons and the colour of the emitted light. For each colour there was a lower bound for velocity, high electron velocities corresponding to colours of low wavelengths. In 1914 James FRANCK and Gustav HERTZ carried out an experiment with colliding electrons under rigorous conditions.[2] Electrons that had traversed a potential of less than 4·9 eV transmitted only momentum to mercury atoms. At higher velocities of the electrons, another process begins to take place. FRANCK and HERTZ assumed that it was the ionization of the mercury atoms. According to EINSTEIN's formula, $E = h\nu$, an energy of 4·9 eV corresponded to the known spectral line $\lambda = 2537$ Å of mercury. Not long afterwards FRANCK and HERTZ were also able to show that this line did in fact occur. According to their explanation, a collision between electrons at energies greater than 4·9 eV could lead to the ionization of the mercury atoms or to the transmission of line 2537. Thus it was that FRANCK and HERTZ had, in 1914, measured an atomic quantity by means of the kinetic energy of electrons (even if they did interpret it wrongly—it was actually the excitation energy), and they were able to prove the validity of the formula $E = h\nu$ for a spectral line.

X-rays

As quantum phenomena are particularly striking for short-wave light, it was important to be able to carry out experiments with X-rays. These had been discovered in 1896 and were generated by the deceleration of cathode rays. They were soon interpreted as electromagnetic radiation of imprecise frequency caused by the deceleration of the electrically charged electrons. They had the property that the individual wave packets—the 'electromagnetic pulses'—were quite short. It was discovered that their ability to penetrate matter, their 'hardness', was greater the higher the energy per decelerated electron. In 1903 J. J. THOMSON concluded from calculations of the scattering and absorption of X-rays by matter that their hardness went hand in hand with what amounted to a kind of effective vibration frequency of the X-ray radiation. This meant that there must also be a direct connection between the magnitude of the (deceleration) energy per electron and the frequency of the radiation thus generated.

When X-rays passed through matter, electrons were released: the 'secondary electrons'. Even for low intensities of X-ray radiation this release set in immediately, so that WIEN was forced to say, in 1905, that the energy of the secondary electrons could not come from the X-ray wave. In that case the energy must surely be distributed inhomogeneously over the wave. The same difficulty had thus emerged as in the case of the release of electrons by ordinary light.

After the appearance of EINSTEIN's 1905 paper introducing the light quantum hypothesis, STARK, in 1907, rather casually, and actually in a footnote, explained how X-rays were emitted as a result of the deceleration of a cathode ray.[3] He put the deceleration time (which he took to correspond to half a wavelength) equal to the quantity h/E which had the dimension of time. E is the decelerated energy. Thus he wrote

$$\lambda/2c = h/E$$

which we may prefer to read as $E = 2h\nu$. He dealt with the appearance of secondary electrons by means of the equation

$$E = hc/\lambda$$

which is equivalent to $E = h\nu$. He did not quote EINSTEIN's paper and would therefore appear to have discovered the relationship independently. In a paper that he wrote a little later, he does mention EINSTEIN and assumes the single formula $E = h\nu$ for both *Bremsstrahlung* and for secondary electrons.[4] In 1907 WIEN assumed the same equation for the generation of X-rays, in common with EINSTEIN, and he used it to determine the wavelength of X-rays.[5] Quantum theory was now also being taken seriously by experimental physicists.

At that time radioactivity also belonged under the heading 'the conduction of electricity by gases'. This phenomenon, discovered at roughly the same time as X-rays, became particularly important for early quantum theory as α-particles turned out to be ideal for probing the inner structure of the atom.

Light Quanta

In the years 1902–7 EINSTEIN completed a basis for statistical thermodynamics which clarified the work of BOLTZMANN and MAXWELL and brought out the essential points. His work was similar to that published by GIBBS in 1902. The temperature of a physical system appeared in the expression for the statistical frequency $\sim \exp\left(-E/kT\right)$ with which a state with energy E arose, and the entropy was given by $S = k \ln W$. This definition of entropy was assumed by EINSTEIN to hold even for non-equilibrium states and he obtained the value of $-\mathrm{d}^2 S/\mathrm{d}x^2$, which gives the sharpness of the entropy maximum as a measure of the reciprocal of the statistical variation of a quantity x

$$\overline{(x - \bar{x})^2} = -1 \left/ \frac{\mathrm{d}^2(\ln W)}{\mathrm{d}x^2}\right. . \tag{3}$$

The greater this statistical deviation the smoother the entropy maximum corresponding to the equilibrium state. Thus EINSTEIN was able to give an explanation of, for example, the Brownian motion of suspended particles in 1905. His analysis of WIEN's and later PLANCK's radiation formulae fitted into this work (1905 and 1909).

The entropy that he calculated in 1905 from WIEN's formula $w \sim \nu^3 \exp\left(-h\nu/kT\right)$ according to

$$1/T = \mathrm{d}s/\mathrm{d}w$$

(which is analogous to $1/T = \mathrm{d}S/\mathrm{d}E$) behaves like the entropy of a gas which consists of independent particles of energy $h\nu$. The entropy $s(T,v)$ is measured per unit volume and unit frequency interval. *Where* WIEN's *formula holds, the thermodynamic behaviour of radiation is as though it consisted of quanta of energy* hv (actually he wrote $R\beta v/N$). EINSTEIN called this an 'heuristic approach to the production and transformation of light', and thus suggested that it was an approach that might well lead to new discoveries. He found confirmation of his light quantum hypothesis in STOKES's law, which states that in photoluminescence the frequency of the light emitted is less than or at most equal to that of the incident light. The photoelectric effect—i.e., the release of electrons from metal surfaces by incident light—gave further confirmation. He wrote an energy equation for this process that was equivalent to equation (2). For the ionization of atoms by light he formulated

$$h v > I$$

(where I is the ionization energy). At that time it was possible to verify the equations with only limited accuracy. The equation $E = hv$ was soon applied by STARK and WIEN, as we have already said, to X-rays, where the quantum phenomena are much more noticeable because of the high frequencies involved.

Relation (2) for the kinetic energy of the electrons in terms of the electric potential U available for retarding the electrons

$$h v = e U + P$$

could be extended to give a method for the precise determination of h/e. That was a laborious task, though, and it was not until 1914 and 1916 that R. A. MILLIKAN was able to make precise measurements.

In 1906 EINSTEIN saw that PLANCK's theory of black-body radiation meant that PLANCK had introduced the hypothesis of light quanta. But EINSTEIN did not yet quite understand what was going on, because a gas composed of light quanta behaved according to WIEN's and not PLANCK's law, or so EINSTEIN's own results told him.

The Wave-Particle Duality for Light

EINSTEIN's 1909 investigation of PLANCK's radiation formula gave a much better understanding of the relationship between light particles and light waves. As he considered PLANCK's method of counting events in the determination of the entropy $S = k \ln W$ for black-body radiation to be unjustified, he reversed the question: what are the consequences for W of PLANCK's radiation formula, which had, after all, been verified empirically? He was able to derive $s(T,v)$, the entropy per unit volume and per unit frequency interval, from $w(T,v)$. The intensity of the fluctuation of the radiation energy within a region of space was then given by $1 : d^2S/dE^2$, according to (3). We recall that PLANCK had originally guessed his radiation formula by interpolation, writing this very quantity as the sum of two terms. The one term on its own would have given WIEN's formula which EINSTEIN explained in terms of the light quantum concept. The other term taken alone would have given RAYLEIGH's formula which follows from the classical concept of light waves. Thus it is that EINSTEIN obtained an expression for the magnitude of the fluctuation, the first term of which gave the energy fluctuation of a gas composed of independent light particles, and the second term of which gave the fluctuation of the energy of a system of classical light waves. EINSTEIN's result was equivalent to the formula

$$\frac{\overline{(E-\bar{E})^2}}{\bar{E}^2} = \frac{1}{z} + \frac{1}{q} \tag{4}$$

where E denotes the energy in a given volume, z the average number of characteristic electromagnetic vibrations and q that of the light quanta it contains. Thus in some way radiation contains not only classical electromagnetic oscillations, but also light particles. EINSTEIN gave only an order-of-magnitude argument for the first term in (4). In 1912 LORENTZ provided the exact calculation.[7]

Now let us consider the relationship between EINSTEIN's result and PLANCK's ideas of 1900. When he made his interpolation PLANCK had constructed the reciprocal of d^2S/dE^2 out of two simple terms added together, the first of which gave RAYLEIGH's and the second WIEN's radiation formulae. According to EINSTEIN, the first terms represented the fluctuation of the classical waves and the second that of the light particles. PLANCK's quantum theory involved a departure from the idea of a continuum of energy values. And this was responsible for the difference from RAYLEIGH's formula. According to EINSTEIN apart from the fluctuation term $1/z$ for the wave there must also be a contribution to the fluctuation from the particles. PLANCK's theory also involved a new method of distributing energy quanta $\varepsilon = h\nu$ (corresponding to what was later to be called BOSE statistics). And this was responsible for the difference from WIEN's formula. According to EINSTEIN, apart from the fluctuation term $1/q$ for the particles, there must also be a contribution to the fluctuation from the waves.

EINSTEIN also examined the vibration of a plate free to move under the action of radiation (analogous to molecular Brownian motion) caused by the fluctuations in the momentum which are related to the radiation pressure. There, too, he found two contributions to the fluctuation, one classical, due to the electromagnetic field, and one derived from the momentum $h\nu/c$ of the light particles. The latter represented a strange but well-defined motion of the light particles—later referred to as '*Nadelstrahlung*' (needle radiation).

With his paper on statistical fluctuation in radiation, EINSTEIN had introduced the wave-particle duality for light. He did not yet, of course, realize the full significance of this revolutionary idea. It was not properly understood until the introduction of the corresponding duality for matter by DE BROGLIE and after the formulation of a new rigorous version of quantum mechanics.

During the discussion at the Salzburg meeting (1909)[8] EINSTEIN said that he expected the fusion of the wave and particle theories of light. PLANCK characterized the difference between his and EINSTEIN's versions roughly as follows: EINSTEIN had made the vacuum the centre of quantum theory, while PLANCK regarded the resonator energies $E = h\nu n$ as paramount. All were agreed that the interaction of radiation and matter must be recognized as the source of the difficulties. The first phase of quantum theory saw the elementary quantum of action as a part of statistical mechanics; the second phase saw it as part of the interaction of light and matter. It was not until much later that its far wider significance was recognized. In fact, it involved the modification of the whole classical picture—even that of motion in which it was not necessary to take account of radiation.

EINSTEIN's light quanta were not taken all that seriously by his contemporaries. STARK alone campaigned vigorously in favour of this concept, emphasizing its plausibility. For example, in 1909 he used the vector combination of the momenta of electrons and the momenta $h\nu/c$ of light quanta[9] when dealing with the deceleration of electrons. PLANCK and SOMMERFELD regarded light

quanta as an unnecessary circumscription of the quantum hypothesis. PLANCK was particularly opposed to the idea. He was then
working to heal the breach between his theory and earlier physics.
His attempts to change quantum theory did not, however, influence its development. Even when EINSTEIN was proposed for
membership of the Berlin Academy in 1913 it was said of him
that he had 'gone too far' with his light quantum hypothesis.
BOHR was still reluctant to accept it in 1922. The hostility or
indifference of physicists did not abate until the discovery of the
COMPTON effect. In 1922 Arthur Holly COMPTON discovered that
an increase in wavelength $\Delta\lambda \sim (1 - \cos \delta)$ arose when X-rays
collided with matter of low atomic weight, the constant of
proportionality being universal. Both COMPTON and DEBYE
showed that the reduction in the frequency corresponded precisely to the reduction in the energy of light corpuscles when they
collided with an electron that could be regarded as free (for low
atomic weight, the force which holds the electron in the atom can
be neglected).[10] It was necessary to apply only the laws of momentum and energy, just as for the collision of ordinary particles.
It follows from the conservation of momentum (for non-relativistic velocity of the electron, see Figure 4) that:

$$m^2 v^2 = (h^2/c^2)(v_0^2 + v^2 - 2v_0 v \cos \vartheta)$$

so that if $v_0 - v$ is not too large:

$$mv^2 = (2h^2/mc^2)v_0 v(1 - \cos \vartheta)$$

the conservation of energy gives:

$$mv^2 = 2h(v_0 - v)$$

and if we compare the two expressions we obtain

$$\Delta\lambda = (c/v) - (c/v_0) = c(v_0 - v)/vv_0 = (h/mc)(1 - \cos \vartheta)$$

for the change in wavelength.

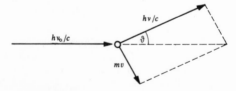

FIGURE 4:   THE COMPTON EFFECT

Light Quanta

It was precisely this change in wavelength that COMPTON measured. It is thus possible to use the ordinary laws of impact for light particles.

In the derivation given by EINSTEIN[11] in 1916 and 1917 he showed that PLANCK's radiation formula had a far more general foundation than the assumption $E(n) = hvn$ for oscillators. It was based on the assumption that light could be emitted or absorbed only in amounts hv and that an atomic system with a higher energy E_2 could spontaneously go over into one with lower energy $E_1 (hv = E_2 - E_1)$ with a certain probability A_{21}. Further transitions proportional to the energy density w of the radiation arose between the two energies with probability coefficients B_{21} and B_{12}. The final assumption was that classical physics was valid for small values of v. This was applied in 1916 in a weaker version to be strengthened in 1917.

The stationary nature of the equilibrium state requires that

$$(A_{21} + B_{21}w)N_2 = B_{12}wN_1$$

where N_2 and N_1 are the numbers of atoms in states E_2 and E_1. So we have for the energy density of the radiation:

$$w = A_{21}/[B_{12}(N_1/N_2) - B_{21}].$$

As
$$N_1/N_2 = \exp(hv/kT)$$

in thermal equilibrium, it follows that

$$w = A_{21}/[B_{12}\exp(hv/kT) - B_{21}].$$

For small v (1917 version), the classical radiation formula

$$w = (8\pi/c^3)v^2kT$$

must hold. It follows that $B_{12} = B_{21}$ and $A_{21} = 8\pi hv^3/c^3$, which gives PLANCK's formula.

[1] A. HERMANN in the Introduction to Vol. 7 of Dokumente der Naturwissenschaft, 1965

[2] J. FRANCK and G. HERTZ, Verh. D. Phys. Ges. **16**, 457, 512 (1914)

[3] J. STARK, Phys. Z. **8**, 881 (1907)

[4] J. STARK, Phys. Z. **8**, 913 (1907)

[5] W. WIEN, Gött. Nachr. **1907**, 598

[6] A. EINSTEIN, Phys. Z. **10**, 185 (1909)

[7] H. A. LORENTZ, Les théories statistiques en thermodynamique, Leipzig and Berlin 1916, p. 114

[8] A. EINSTEIN, Phys. Z. **10**, 817 (1909)

[9] J. STARK, Phys. Z. **10**, 902 (1909)

[10] A. H. COMPTON, Phys. Rev. **21**, 483 (1923)
P. DEBYE, Phys. Z. **24**, 161 (1923)

[11] A. EINSTEIN, Verh. D. Phys. Ges. **18**, 318 (1916), Phys. Z. **18**, 121 (1917)
(English version in B. L. van der WAERDEN's Sources of Quantum Mechanics)

4. THE ATOM

The Problem of the Atom

QUANTUM theory started life as quantum statistics. It was later to become the problem of the interaction of light and matter. It is, however, more fashionable today to see the core of quantum theory as consisting in the fact that it enables us to understand the atom. The atom is where we see quantum theory in action. Non-intuitive in nature, quantum mechanics is valid when we are dealing with atomic orders of magnitude. Classical mechanics, which is much easier to visualize, holds for macroscopic dimensions. Because of this, atomic research made a fundamental contribution to the understanding of quantum theory, if not to the actual discovery of the elementary quantum of action. From the very beginning, there was a special theoretical problem concerning the atom. The atom was required to be the ultimate indivisible component of matter, without which matter seemed to be incomprehensible. But it had to be thought of with a definite extent and a certain structure. So it was, at least in principle, divisible. KANT's second example of the antinomy of pure reason dealt with this point. Nowadays we have a solution to this problem. For us the atom is the ultimate component of matter in the intuitive sense of the word. If, on the other hand, we are talking about the structure of the atom itself, we are no longer using the word in its intuitive sense. But we have been able to speak in this way only since the 1920s. The theory of specific heats gives us an indication of the situation. Atoms and molecules conceal their inner nature. They behave, to a large extent, like particles or rigid systems of particles, especially at low temperatures.

The Chemical Atom

The idea that matter consisted of atoms was an attempt to solve an age-old problem, the problem of matter and its peculiar properties. Apart from the problem of change, or more particularly that of motion, and the problem of the cosmos, it was one of the major problems of philosophy and physics. The ancients had provided two solutions to the problem of matter. The first was the theory of the four elements, which was adopted by ARISTOTLE, and which had therefore been used by traditional philosophy. The second was atomism, which postulates the existence of atoms and empty space. Changes in nature are conceived of as the motion of particles through a vacuum. Atomism remained very much in the background, even later on in the Arab world.

The new chemistry of the seventeenth century made use of atomistic concepts. NEWTON's physics was a mechanics of particles in a vacuum. BOYLE saw matter as composed of particles with specific qualitative properties. The development of this chemistry led, through the clarification of the idea of a chemical element by precise weighing, and through the whole-number laws, to the view expressed by DALTON in 1808, that the particles of chemical compounds were aggregates of a definite number of particles of the chemical elements. And thus the problem of matter became that of the atom. Our modern notation for chemical formulae is an expression of the atomic structure of molecules. It is based on the law of mass ratios and the law of volume ratios for gases. The idea of a mole, which lies at the basis of the volume law, and AVOGADRO's law (which states that the volume of a mole of gas is independent of the type of gas and that a mole of gas contains a definite number of molecules) were, however, not established until around 1860.

For later theories of atomic structure the concept of valency was of particular importance. It was understood as soon as chemical formulae were reasonably clear, by around 1860. Soon homologies were discovered among the elements, for example the series Na, K, Rb, Cs or Mg, Ca, Sr, Ba. This culminated in the periodic system of the elements (1868–9). The gaps in this system were gradually filled. The noble gases were added at the end of the nineteenth century. The period lengths 2, 8, 8, 18, 18 were then well established. There was, however, no general idea of what caused valency. In inorganic chemistry the tendency had been (since BERZELIUS) to assume electric charges on the atoms, and to explain affinity and valence numbers in this way. In organic chemistry it was assumed that forces went out from particular valence points of the atoms (LE BEL, VAN'T HOFF).

There were good reasons for both versions, and finally both appeared in the quantum theory of chemical bonds. German chemists were particularly averse to theory. Physicists were also very cool towards such models of atoms. BOLTZMANN did not use AVOGADRO's constant (the number L of molecules per unit volume or per mole) nor the equation $R = Lk$ (for the gas constant). PLANCK also wanted to avoid using the atomic hypothesis. Wilhelm OSTWALD and Ernst MACH were prepared to grant no reality whatsoever to atoms. But finally the physicists managed to measure properties of the atom and L was measured in several different ways, all of which gave the same value. As physicists knew about electrons, the chemical models were ultimately supplanted by those of the physicists.

Electrons within the Atom

The electron was conceived first as a unit of charge for electrolysis, then for cathode rays and for emission from the surface of metals, and finally as an emitting electron in the line spectra of the atoms. AVOGADRO's constant L is connected with the FARADAY constant F, the charge transferred by electrolysis per equivalent, by the formula

$$eL = F.$$

Within the atom there are electric particles with charge $\pm e$. The valency of the element is connected with this. It was possible to measure e/m for a cathode ray. The particles of negative charge that can be released from atoms have, according to this, a very small mass. It appears to be impossible to release positive charge from the atom. The atom thus consists of positive electric charge and of electrons. The electrons within the atom were held responsible for the spectral frequencies in the visible and ultraviolet region. In 1897 H. A. LORENTZ was able to explain the splitting of spectral lines which had just been discovered by Pieter ZEEMAN in an external magnetic field. By comparing the equation of motion of an electron in a magnetic field $\boldsymbol{B}$

$$m\dot{v} = -ev \times \boldsymbol{B}$$

with the equation of motion of a particle in a rotating system of reference (with angular velocity ω)

$$m\dot{v} = 2mv \times \omega$$

we see that the magnetic field imparts to the electrons an additional rotational motion such that

$$\omega = \frac{e\,B}{2m}.$$ (1)

This gives precisely the ZEEMAN splitting for typical cases. Its measurement is, as shown by (1), at the same time a determination of e/m. LORENTZ calculated from ZEEMAN's measurements the same value for e/m as was found for cathode rays. The electrons in an atom are the same as those in a cathode ray. The spectral lines derive from the motion of these electrons. This sums up the views of physicists concerning the atom around the turn of the century. Well, how many electrons were there in an atom? From the complicated structure of the spectra it was often concluded that there were a large number of them. However, in 1903 J. J. THOMSON worked out from the intensity of scattering of X-rays that the number of electrons must be roughly equal to the atomic weight.

Models of the Atom

Thus the atom consisted of a positively charged component and a number of electrons. Very little was actually known about the positively charged part of the atom but it was always possible to speculate about it. Thus it was that shortly after 1900 models of the atom were suggested with a point-like positive nucleus, and some with uniformly distributed positive matter.[1] Nuclear types of atom included the 'nucleoplanetary structure' of Jean B. PERRIN (1901) and the 'Saturnian system' of H. NAGAOKA (1904). KELVIN (1902) suggested a model with a uniformly distributed positive charge and considered the stability of systems of electrons. However, he was unable to explain the periodic system of the elements. In 1903 J. J. THOMSON extended the model. 'As we know nothing about the positive charge' it was reasonable to distribute it uniformly over a sphere. In this field there were arrangements for the equilibrium of 1, 2, 3,... electrons. For example, for four electrons there was a tetrahedral arrangement. As THOMSON was unable to find a general solution for N electrons, he made a simplified version of his model by arranging the electrons regularly in plane rings. There could be up to five electrons in any one ring. With more rings it was possible to accommodate more electrons in the second, third, etc., rings (counting outwards). The maximum number of electrons that could lie in the rings were 5, 11, 15, 17, 21, 24. By calculation and experiments with floating magnets he discovered a stable distribution among the rings for each number of electrons. Table 1 gives an extract from his results (the occupation numbers of the rings are given counting from the innermost one).

TABLE 1: J. J. THOMSON'S RINGS OF ELECTRONS

5	5	16	5+11	57	2+8+12+16+19
6	1+5	17	1+5+11	58	2+8+13+16+19
7	1+6	18	1+6+11	59	2+8+13+16+20
8	1+7	19	1+7+11	60	3+8+13+16+20
9	1+8	20	1+7+12	61	3+9+13+16+20
10	2+8	21	1+8+12	...	...
...		22	2+8+12	66	5+10+14+17+20
		23	2+8+13	67	5+10+15+17+20
		...		68	5+10+15+17+21
				69	5+11+15+17+21
				70	1+5+11+15+17+21
				...	...

It was now reasonable to expect that arrangements that had the same occupation numbers in certain rings should display similar physical and chemical properties, for example the arrangements of 5 and 16 electrons, 6 and 17 and so on, shown side by side in the table. Thus THOMSON gave a model (it was not intended to be any more than this) for the chemical homology of elements. If for example one considered all the arrangements that had twenty electrons in the outermost ring one might well regard the one with a total of 59 electrons as not particularly stable, with the result that this arrangement might easily give up an electron as a result of some kind of disturbance. This would therefore be analogous to a strongly electrically positive element. A corresponding result held for the arrangement with a total of 68 electrons. Thus THOMSON also derived a model for chemical valency. Certainly valency and homology were not as closely linked as in the periodic system of elements, but THOMSON was, after all, considering only a plane model of what was in fact a three-dimensional reality. STARK had very detailed if somewhat fantastic ideas about the positively charged part of the atom (1907 onwards).[2]

Soon after 1900 it was recognized that the fundamental characteristic of the atom of a chemical element was the number of electrons, which was sometimes set equal to the atomic weight, occasionally less, about half as large, so that there was one chemical element for each number. This opened up the tempting prospect of describing a chemical element completely in terms of a single number and of being able to derive its chemical and physical properties with the aid of this one number by purely theoretical means. In models of the THOMSON type the stability of the atom was not as yet a serious problem. Electrons in a positive volume distribution of charge have stable positions. Spectra, however, were still not understood at all.

Experimental research decided the arrangement of the positive charge in favour of the nuclear type of atom. While the models of PERRIN, THOMSON and NAGAOKA were purely speculative, measurements of the way very fast particles generated by radioactive decay travelled through matter gave more precise indications. LENARD investigated the absorption of β-rays and in 1903 drew the conclusion from his measurements that an atom was permeable for sufficiently fast particles and that its interior was a force field rather than a distribution of matter. In RUTHERFORD's laboratory, the transmission of α-particles was being investigated. In 1909 Hans GEIGER and Ernest MARSDEN discovered that some of the α-particles experienced a very strong deflection and must therefore be exposed to a strong force field. RUTHERFORD now calculated the deflection of particles by a strong central point charge around which the compensating charge could be distributed uniformly. He found it to be in accord with experiments.[3] Even if he was not thus able to determine the sign of the central charge, it seemed fairly clear that this must be positive. RUTHERFORD therefore concluded that the positive charge and most of the mass of an atom was concentrated in a very small space and thus that an atom was made of a practically point-like central nucleus and of electrons. He was able to estimate the number of the elementary charges in the nucleus as about half the atomic weight, using the measurements of GEIGER and MARSDEN. RUTHERFORD's result was supported by the proof in 1912 of sharp bends in the traces of α-particles by C. T. R. WILSON in his cloud chamber. Further confirmation came in 1913 from more precise measurements made by GEIGER and MARSDEN which in particular verified the charge number of the nucleus as being roughly half the atomic weight.

The nuclear charge number was now the only parameter that could determine the properties of chemical elements (apart from the atomic weight). It would have been difficult to avoid noticing that it corresponded to the atomic number of the element, the number in the sequence of the periodic system (although this did differ very slightly from the sequence of atomic weights). But absolute certainty did not come until the radioactive elements had been brought into the periodic system and it had been seen that the element moved around two points when transmitting α-particles, while two positive elementary charges (a He^{++}-ion) were emitted. In 1912 Antonius van den BROEK put the atomic number equal to the charge number of the nucleus.[4]

Spectral Series[5]

Striking, seemingly regular sequences of lines arise in the spectrum of many elements. The first lines of such a sequence for the hydrogen spectrum were represented by BALMER in 1885 by the formula

$$\lambda = \frac{Cn^2}{n^2-4}$$

which was later written as

$$\frac{1}{\lambda} = R\left(\frac{1}{4} - \frac{1}{n^2}\right).$$

The fact that there are similarities between many spectra if they are treated in terms of $1/\lambda$ rather than λ was noted by W. HARTLEY in 1883. In 1888 recurrent $1/\lambda$-differences were observed in a number of spectra. RUNGE wrote the BALMER series for hydrogen in 1888 as:

$$\frac{1}{\lambda} = A + \frac{B}{n^2} \ (B<0)$$

other series as:

$$\frac{1}{\lambda} = A + \frac{B}{n^2} + \frac{C}{n^4}$$

and in a few rare cases:

$$\frac{1}{\lambda} = A + \frac{B}{n} + \frac{C}{n^2} + \cdots$$

The theory developed in two ways. Once H. A. ROWLAND had developed high-resolution spectral gratings it was possible to begin exact measurements of the spectral series. Thus it was that KAYSER and RUNGE were able to carry out a precise scrutiny of the spectra of Na, K, Rb, Cs, of Mg, Ca, Sr, Ba, Zn, Cd, Hg, of Cu, Ag, Au, of Al, In, Tl, of Sn and Pb, of As, Sb and Bi, in the years 1888 to 1892. RUNGE and PASCHEN were able to add series for O, S, Se and He in 1895–7. They discovered the two spectral systems of orthohelium and parahelium in the process. The second way in which the theory developed was as a result of the introduction by RYDBERG in 1889 of series formulae, based on available material, which expressed the fundamental idea more satisfactorily:[6]

$$\frac{1}{\lambda} = A - \frac{R}{(n+\alpha)^2}.$$

He saw also in the case of alkaline spectra that A was equal to the second term of another series, so that he was also able to write

$$\frac{1}{\lambda} = \frac{R}{(n_1+\alpha_1)^2} - \frac{R}{(n_2+\alpha_2)^2}. \tag{2}$$

He thus wrote a spectral frequency as the difference between two 'terms'. He recognized the series terms s, p, d for alkaline spectra (and with slight deviations also for other spectra), with the combination terms s–p (primary series), p–s (sharp), p–d (diffuse secondary series). On closer investigation spectral lines frequently turned out to consist of very closely-packed groups of lines so that the corresponding terms could be simple, double (doublets) or triple (triplets). Already in 1893 RYDBERG had written the frequencies of the combination of two triplet terms in a square array of three rows and three columns with empty places.[7] He thus derived what was later to become the principle of the combination of spectra.

It was perhaps a disadvantage in the history of quantum theory that this fundamental spectral law was first given in a form which was too closely connected with the overspecialized expression (2). Little attention was paid to this. KAYSER and RUNGE were very reluctant to adopt (2): after all α was not exactly constant even within a given series, and R varied slightly from element to element. But around 1900 tables of spectral lines were quite often written in a rectangular arrangement. It was not until 1908 that the general form of the combination principle was given by RITZ[8]:

$$1/\lambda = F(n_1,\alpha_1,...) - F(n_2,\alpha_2,...) \tag{3}$$

In this form it holds exactly. But of course not all λ that can be written in this way necessarily arise. PASCHEN immediately applied RITZ's form of the principle and in particular discovered a number of lines that it predicted for infra-red rays, and also the series for hydrogen

$$\frac{1}{\lambda} = R \left(\frac{1}{9} - \frac{1}{n^2}\right).$$

Several attempts were made to give a theoretical interpretation of the spectral series. For example, an attempt was made to deal with the form of the series formulae expressed in terms of differences by using the harmonics that arise in non-linear wave equations. Elastic systems were also constructed with frequencies that had finite points of accumulation. RITZ made a number of attempts of this kind. In doing so in 1903 he discovered the following formula[9]:

$$\frac{1}{\lambda} = A - R/\left[n + \alpha + \beta\left(A - \frac{1}{\lambda}\right)\right]^2$$

which is also very useful for the representation of empirical spectra. But the attempts at theoretical interpretation of the spectra were soon felt to be rather unsatisfactory.

h *and the Atom*

The fact that the quantum of action h arose in the interaction between light and matter suggested the idea of trying to explain the atom in terms of h—or, conversely, to explain h in terms of the atom.

STARK provides an example of this approach. He was a very good experimental scientist. In 1905 he discovered the DOPPLER effect for canal rays and made significant use of it. In 1913 he discovered the splitting of the hydrogen lines in an external electric field, the 'STARK effect' for which he was awarded the Nobel prize. Open-minded and not in the least prejudiced by current thinking, gifted with a vivid imagination which reached out far beyond the realm of what could be proved, STARK attempted to answer those questions that were later to be solved by BOHR—the problems of the spectra and of chemical bonds. But after BOHR's success it was no longer worth paying much attention to STARK's theory of valency (1908) nor to his principles of atomic dynamics (1910, 1911, 1915).[10] According to STARK, the band spectra and line spectra were connected with the ionization of atoms. Ionization occurred as a result of the collision of electrons, for which there was a lower threshold value of the kinetic energy (1902). From 1907 STARK recognized the importance of the formula $E = h\nu$ for ionization and for spectra. He did not, however, apply the series laws and his ideas did not bear fruit. From 1910 he distinguished emphatically between 'atomic dynamics' for atomic quantities and the 'dynamics of media' which was valid wherever the atomic nature of matter could be ignored. This, however, did not prevent him from sketching in his theory of valence extremely vivid pictures of atoms with an extensive positive sphere and electrons.

While STARK was unable to give the precise connection between h and the atom, Arthur Erich HAAS (1910) succeeded in formulating a quantitative relationship between h and the radius of an atom.[11] In the THOMSON model of the hydrogen atom the single electron moves round a homogeneous positive volume distribution of charge. Within this the force field is that of an harmonic oscillator. Up to the boundary $r = a$ the force is e^2r/a^3, the potential energy $e^2r^2/2a^3$ (calculated with respect to the centre), the total energy twice this value. At the boundary itself the energy is thus equal to e^2/a. For an electron orbiting round the boundary HAAS, using ordinary mechanics, put

$$m\omega^2 a = e^2/a^2.$$ (4)

Using quantum theory, he had to put

$$h\omega/2\pi = e^2/a.$$ (5)

Together these gave

$$(h/2\pi)^2 = e^2ma.$$ (6)

The formula remains true in BOHR's theory, while (4) is also valid for the RUTHERFORD model of the atom and (5) becomes $h\nu = 2\overline{E_{\text{kin}}}$ in BOHR's theory. By equating the frequency in (4) and (5) to the limiting frequency of the BALMER series $\omega/2\pi = Rc/4$—the LYMAN series $R(1 - 1/n^2)$ was not yet known—HAAS obtained values for e and a that no longer held in the later improved theory. Order-of-magnitude considerations by A. SCHIDLOFF (1911)[12] pointed in the same direction but they were less successful. J. W. NICHOLSON in 1912 put the angular momentum of electron orbits equal to $h/2\pi$, but used this rather uncritically in connection with stellar spectra.[13]

To give some idea of the attitude of physicists to the problem of the atom around the years 1910–1, we must first say that STARK recognized that the dynamics of atoms was very different from that of macroscopic phenomena. PLANCK saw the difference between physics and chemistry as consisting in the fact that the latter was based on quantum theory. SOMMERFELD expressed his own attitude at a meeting in 1911: *some people think that the elementary quantum of action is in some way a consequence of the atomic structure of matter; but that is not so, h is much rather a key to the understanding of the atom.*[14] This turned out to be true. The man who turned the key was Niels BOHR, above all by recognizing the connection between quantum theory and the combination principle of spectroscopy. This connection, in which the combination principle

$$v = F(n_1 ...) - F(n_2 ...)$$

was merely a special case of the formula

$$hv = E(n_1 ...) - E(n_2 ...)$$

was not recognized until 1913. In Chapter 2 we saw where HASENÖHRL went wrong without it.

[1] L. ROSENFELD, Introduction to new edition of Bohr's work on the constitution of atoms and molecules, Copenhagen 1963

A. HERMANN, Introduction to Vol. 5 of Dokumente d. Naturwiss. 1964

J. PERRIN, Rev. Scientific **15**, 449 (1901)

H. NAGAOKA, Phil. Mag. **7**, 445 (1904)

LORD KELVIN, Phil. Mag. **3**, 257 (1902)

J. J. THOMSON, Electricity and matter 1903

[2] e.g. B. J. STARK, Jb. Radioakt. u. Elektronik **5**, 124 (1908)

[3] E. RUTHERFORD, Phil. Mag. **21**, 669 (1911)

[4] A. VAN DEN BROEK, Phys. Z. **14**, 32 (1913)

[5] H. KAYSER, Handb. d. Spektroskopie Vol. 1 (1900)

[6] J. R. RYDBERG, K. Svenska Vetensk. Ak. Handl. **23**, Nr. 11 (1890), Phil. Mag. **29**, 331 (1890)

[7] J. R. RYDBERG, Ann. d. Phys. **50**, 629 (1893)

[8] W. RITZ, Phys. Z. **9**, 521 (1908)

[9] W. RITZ, Diss, Göttingen 1903, Oeuvres p. 1.

[10] J. STARK, Jb. Radioakt. u. Elektronik **5**, 124 (1908), Prinzipien der Atomdynamik, 3 Vols., Leipzig 1910–15

cf. A. HERMANN's Introduction to Vol. 7 of Dok. d. Naturwiss

[11] A. E. HAAS, Sitz. Ber. Wiener Akad. **119**, 119 (1910), Jb. Radioakt. u. Elektronik **7**, 261 (1910)

cf. A. HERMANN's Introduction to Vol. 10 of Dok. d. Naturwiss. 1965

[12] A. SCHIDLOFF, Ann. d. Phys. **35**, 90 (1911)

[13] J. W. NICHOLSON, Monthly Not. Astr. **72**, 49, 139, 677, 693 (1912)

[14] A. SOMMERFELD, Phys. Z. **12**, 1057 (1911)

5. NIELS BOHR, 1913

The first Paper on the Structure of the Atom

THE next major step in quantum theory after PLANCK's introduction of energy quanta and EINSTEIN's hypothesis of light
quanta was BOHR's first paper of 1913. It was completed in April
and appeared in July. In the same year DEBYE had asserted the
general validity of $\oint p \, \mathrm{d}x = hn$ and EHRENFEST had derived the
quantum states of a rotator. Although quantum theory had
hitherto consisted almost entirely of quantum statistics, or had
been concerned with the incomprehensible features of the interaction of light and matter, thanks to BOHR it now became the
dynamics of the atom. He established the connection between
the states of atoms and the spectral series and he went on to give
a method of calculating the energies of the states of an atom—the
powerful method that was later to be known as the correspondence
principle.

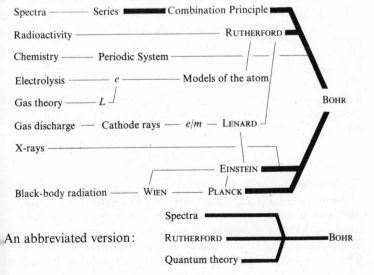

He saw that quantum theory signified a radical break with the philosophy of classical physics and he showed one way of involving classical physics in the construction of the new theory. He noted that the stability of atoms as it emerged in chemistry and physics (stability with respect to collision and radiation) required a new principle for its comprehension and that quantum theory was that principle.

BOHR's principal ideas are still of considerable validity today. We can now represent the lines of development that led to them in the diagram on page 65. This should be quite easy to understand in the light of what we have said: it gives a more detailed version of part of Figure 1.

The first of BOHR's 1913 papers 'On the Constitution of Atoms and Molecules I'[1] assumes the RUTHERFORD model of the atom. As BOHR says in his introduction, in the THOMSON model the radius was given, and the stability of the electron configuration could, in principle, be understood as that of an equilibrium arrangement. However, the theoretical investigation of the new model by means of classical mechanics and electrodynamics does not lead to any understanding of the stability of an atom or of how to determine its radius. It is in fact impossible to construct a quantity with the dimension of length out of the quantities e and m which appear in the model. But, on the other hand, recent experimental and theoretical developments had shown that classical electrodynamics was not really capable of describing systems of atomic size. The introduction of the quantity h transforms the question of stability, and it is perfectly possible to construct a length of the order of the radius of the atom using the quantities h, e and m.

In the first part of his first paper BOHR considered an atom with nuclear charge Ze and with one electron. He derived the correct expression for the energy $E(n)$. It was fortunate that he used as his starting point the relationship

$$v = \sqrt{2}|E|^{3/2}/(\pi Ze^2 \sqrt{m}) \tag{1}$$

between frequency and energy. This equation comes from the well-known mechanics of the motion of an electron in an elliptic path, disregarding radiation. However, although classical physics suggests that the radiation causes the electron to trace a uniformly decreasing orbit we are obliged to assume non-uniform transitions according to quantum theory, as a result of the radiation of an energy $h\nu n$. If we now further assume that while an electron is bound to the nucleus a definite frequency ν is radiated, then we may expect this to be the average of the frequencies zero for the free electron and $\nu(E)$ for the bound electron, $\nu = \nu(E)/2$. We thus obtain for the possible binding energies

$$|E| = \tfrac{1}{2}h\nu(E)n. \tag{2}$$

Using the classical equation (1) for $\nu(E)$ this gives

$$|E| = 2\pi^2 m Z^2 e^4 / h^2 n^2. \tag{3}$$

The firmest bond occurs for $n = 1$. This gives the ground state, which is thus stable. For $n = 1$, $Z = 1$ we obtain values for the semi-axes, the frequency and the binding energy that are of the order of magnitude of the measured radius, of the optical frequency and of the ionization energy of the hydrogen atom. BOHR now sums up his results as follows: in the 'stationary states' classical mechanics is used (i.e., for $\nu(E)$). The transition between two states is accompanied by radiation of a definite frequency. This is the frequency given by quantum theory. BOHR has more to say in the third part of his paper on the subject of the more specialized assumption that the radiation is a multiple of $h\nu(E)/2$.

In the second section the energies $E(n)$ for $Z = 1$ are related to the series of the hydrogen spectrum. During the transition from state n_2 to state n_1 the formula

$$h\nu = E(n_2) - E(n_1) \tag{4}$$

means that the frequency

$$\nu = \frac{2\pi^2 m e^4}{h^3}\left(\frac{1}{n_1^2} - \frac{1}{n_2^2}\right) \tag{5}$$

is radiated. For $n_1 = 2$ this corresponds to the BALMER series, and for $n_1 = 3$ to the PASCHEN series. The common factor agrees well with the observed RYDBERG constant. As the scale of the electron path increases rapidly with n it is possible to understand why so many more lines can be observed of the BALMER series for stellar spectra than in gas discharge tubes. Other spectral series hitherto ascribed to hydrogen with half-integer values of n could now be seen to be lines of the simply ionized helium atom ($Z = 2$):

$$\nu = \frac{2\pi^2 m e^4}{h^3}\left[\frac{1}{\left(\frac{n_1}{2}\right)^2} - \frac{1}{\left(\frac{n_2}{2}\right)^2}\right] = \frac{2\pi^2 m \cdot 4 e^2}{h^3}\left[\frac{1}{n_1^2} - \frac{1}{n_1^2}\right].$$

For atoms with more electrons we have more complicated formulae. But the RYDBERG-RITZ combination principle

$$v = F(n_1 \ldots) - F(n_2 \ldots)$$

holds and we can now see that this is a consequence of (4). If the spectrum arises as a result of the acquisition of an electron, and if this is at a great distance, similar relationships hold to those for the hydrogen atom. We may thus expect to obtain the result:

$$\lim [n^2 F(n \ldots)] = 2\pi^2 m e^4 / h^3$$

which is in accord with RYDBERG's series formulae.

The third part of BOHR's paper is of particular importance. It included a prescription for quantization, which admitted of generalization: this was the seed of the 'correspondence principle'. Assumption (2) is not required. More generally we may take

$$|E| = f(n) h v$$

and from (1) and (4) we obtain:

$$|E| = \frac{\pi^2 m Z^2 e^4}{2 h^2 f^2} \qquad v = \frac{\pi^2 m Z^2 e^4}{2 h^3} \left[\frac{1}{f(n_1)^2} - \frac{1}{f(n_2)^2} \right]. \tag{6}$$

This gives the empirical series formulae for $f \sim n$, let us say $f = cn$. We can determine the value of the factor c by considering the limit for large n. With the difference $n_2 - n_1 = 1$, (6) gives for large n:

$$v = \frac{\pi^2 m Z^2 e^4}{h^3 c^2 n^3} = \frac{2\sqrt{2} c}{\pi Z e^2 \sqrt{m}} |E|^{3/2}.$$

This is formula (1) for classical mechanics if we put $c = \frac{1}{2}, f = n/2$. For $n_2 - n_1 = \tau$ we obtain in the limit for large n another classical frequency, namely the harmonic frequency $\tau v(E)$ that occurs in the motion with fundamental frequency $v(E)$. Thus if we require the quantum frequencies given by (4) to appear in the classical relationship (1) between energy and frequency for large orbits, this in effect determines the quantity $f(n)$. And if we obtain a result as simple as $f = n/2$ for large n we may reasonably assume this also to hold for smaller n. It would not be necessary to refer to the empirical series formula in order to determine f.

BOHR went on to show in the third part that the angular momentum for a circular path was

$$P = (h/2\pi)n$$

and thus $h/2\pi$ for the ground state.

While radiation has been treated up till now as emission, the fourth part deals with absorption, which can lead to excitation and ionization. In the latter case the electron receives a kinetic energy

$$E_{kin} = h\nu - |E|.$$

The fifth part of the first paper was a preparation for Part II, the second paper, in which the chemical properties of the elements were to be explained in terms of rings of electrons.

The major advance made by BOHR's paper can be seen in two points: the RYDBERG-RITZ combination principle

$$\nu = F(n_1 ...) - F(n_2 ...)$$

is a representation of the quantum formula

$$h\nu = E(n_2 ...) - E(n_1 ...)$$

in terms of spectra. The frequency

$$\nu = \frac{1}{h} \left[E(n+\tau) - E(n) \right]$$

tends to the classical frequency $\tau\nu(E)$ for large quantum numbers n. This gives us a method of calculating the energy values. This is at once a departure from classical physics and a bridge between the classical description and quantum behaviour. There are a few more comments we must make on both these points.

Towards the Assimilation of the Spectral Laws

We know quite a lot about how BOHR's paper of the summer of 1913 came to be written[2]. In the summer of 1912 BOHR was carrying out experiments under RUTHERFORD in Manchester, after a short and rather disappointing stay with THOMSON in Cambridge. The aspect of RUTHERFORD's model of the atom that fascinated BOHR was the idea that all physical and chemical properties of a chemical element could be determined by a single number, the number of elementary charges in the nucleus, which is equal to the number of electrons. But he immediately saw that the stability of atoms could not be explained in terms of the RUTHERFORD model. Fragments exist of a first draft dating from the summer of 1912, in which BOHR emphasized that no length was indicated by the properties of the model, that the motion of the electrons was not stable and that the problem of stability should really be dealt with from a completely different point of view. It would be necessary to introduce a relationship $E_{kin} = Kv$ corresponding to the equation given by PLANCK and EINSTEIN. There was no hope of a mechanical explanation. The chemical properties and the periodic system of the elements would have to be determined in terms of rings of electrons, and the electrons in the outermost ring would be decisive. This differed, of course, from J. J. THOMSON's view. BOHR promised, in addition, to explain the periodically changing atomic volume of the elements, the dependence of several properties on the atomic number, and the stability of chemical compounds. No mention was made of optical spectra. It is clear from BOHR's letters that up to February 1913 he had not considered the spectral laws. However, at the beginning of March he sent the first sections of his paper to RUTHERFORD. His accompanying letter shows that the hydrogen spectrum and the value of the RYDBERG constant had been explained. The printed paper is dated 5th April. The observations concerning the He^+ spectrum and perhaps even the formulation of the correspondence principle may have been added after the end of March. So BOHR discovered the theory of the hydrogen spectrum in less than a month. He was later to say that as soon as he had seen BALMER's formula, everything became clear. There are also reports of a conversation between BOHR and H. M. HANSEN, who had just returned from Göttingen in 1913, in the course of which HANSEN asked whether BOHR could explain the spectral laws in terms of his theory. BOHR then said that that would be far too difficult, and HANSEN pointed to the simple RYDBERG laws. As we have said, we can only conclude from his letters that BOHR assimilated the spectral laws into his theory at a very late stage. This is astonishing as BOHR must be assumed to have been in close contact with the University of Lund where RYDBERG had discovered his laws. We can deduce from the rather different relationship between the frequency and the radiated energy in the first and third sections—in the first it was hvn and in the third it was hv—that the first section is derived from an earlier version.

Towards the Correspondence Principle

In order to show what the third section of BOHR's paper contains, we shall change his conclusions slightly. First let us consider two versions that both derive from the classical expression $v(E)$.

The quantum frequency

$$[E(n + \tau) - E(n)]/h$$

corresponds to the frequency $\tau v(E)$ emitted according to classical theory. Let us now require that the two relationships between E and the frequency become more and more similar as n increases. BOHR might well have deduced, slightly at variance with his actual conclusions, that in the Coulomb field we have

$$v(E) = a|E|^{3/2}.$$

In quantum theory we take

$$|E| = h v(E)f(n) = h a|E|^{3/2} f(n)$$

and obtain

$$|E| = 1/(ahf)^2.$$

Thus, for large n the quantum frequency ($\tau = 1$) becomes

$$\frac{\Delta E}{h} \approx \frac{2af'(n)}{(ahf)^3} = 2a|E|^{3/2} f'(n) = 2v(E)f'(n)$$

where the prime denotes differentiation with respect to the given variable. This frequency tends to the classical frequency if we put

$$f'(n) = \tfrac{1}{2} \qquad f = \tfrac{1}{2}(n + c).$$

$c = 0$ corresponds to the series formulae. The generalization for arbitrary $v(E)$ is:

$$E(n) = h v(E)f(n)$$

$$f(n) = \frac{E(n)}{h v(E)}$$

$$f'(n) = \frac{E'(n)}{h v(E)} \left[1 - \frac{E v'(E)}{v(E)} \right].$$

For large values of n, $\Delta E/h$ then tends to $v(E)$ if the factor in front of the brackets is equal to unity:

$$f'(n) = 1 - E \frac{\mathrm{d}}{\mathrm{d}E} [\ln v(E)].$$

It is therefore possible to calculate $E(n)$ from $v(E)$, with an arbitrary constant of integration. For power series we have the particularly simple result $v \sim E^r$. This gives $f' = 1 - r$ and therefore

$$E = (1-r)hvn.$$

For the harmonic oscillator $r = 0$ and $E = hvn$. For the rotator $r = \frac{1}{2}$ and $E = hvn/2$. For the Coulomb field we have $r = 3/2$ and $E = -hvn/2$.

This can all be shortened quite considerably. The frequency $[E(n + \tau) - E(n)]/h$ that is radiated according to quantum theory can be written in the form $\tau \, \mathrm{d}E/h \, \mathrm{d}n$ for large n. In this limit it agrees with the classical frequency $\tau v(E)$, if

$$v(E) = \frac{\mathrm{d}E(n)}{h\,\mathrm{d}n}$$

$$hn = \int \frac{\mathrm{d}E}{v(E)}.$$

This is the HASENÖHRL quantum condition of 1911. In 1914 BOHR gave this form of the working.

The connection with phase extension follows if we recall the theorem in classical mechanics:

$$v = \frac{\mathrm{d}E}{\mathrm{d}\Phi} \qquad \Phi = \oint p\,\mathrm{d}x = \frac{2\overline{E_{\mathrm{kin}}}}{v}.$$

The classical frequency $\mathrm{d}E/\mathrm{d}\Phi$ then appears as the limiting value of the quantum-theoretical frequency $\Delta E/\Delta \Phi$ where

$$\Delta \Phi = h.$$

BOHR knew this form too, in 1914. We may apply the method

$$2\overline{E_{\mathrm{kin}}} = hv(E)n$$

according to BOHR, quite generally, as

$$\mathrm{d}E = v\,\mathrm{d}\left(\frac{2\overline{E_{\mathrm{kin}}}}{v}\right) = hv\,\mathrm{d}n$$

tends asymptotically to

$$\Delta E = h\nu$$

for large n.

Further Papers

In the lecture that BOHR gave in December 1913 to the Physics Society in Copenhagen[4] he emphasized the asymptotic agreement between classical and quantum-theoretical frequencies. For the hydrogen spectrum we may deduce from BALMER's formula, putting $h\nu = \Delta E$, that

$$E(n) = -\frac{Rhc}{n^2}.$$

The classical expression

$$\nu = A|E|^{3/2} \tag{7}$$

would seem to imply something quite different. But for large n the quantum frequency ($\tau = 1$) becomes

$$\nu = \frac{2Rc}{n^3} = B|E|^{3/2}.$$

The two coincide in the limit if $A = B$. This gives an expression for the RYDBERG number R in terms of e, m and h. Strictly speaking, we should use some average value of the nuclear mass and that of the electron in expression (7). This means that

$$\nu \sim \left(\frac{1}{m} + \frac{1}{M}\right)^{1/2} |E|^{3/2}$$

which explains the slight difference in the RYDBERG numbers for hydrogen, helium and the heavy elements.

In the second part of his 'trilogy' on the structure of atoms and of molecules, which appeared in September 1913, BOHR attempted to explain spectroscopic and chemical properties in terms of rings of electrons. The stabilization of the rings arises as a result of the condition that the angular momentum be $h/2\pi$ per electron. BOHR's imagination somewhat outran the possibility of definite proof. These observations mark the beginning of BOHR's explanation of the periodic system of the elements. We shall deal with this in Chapter 8. The third part, which appeared in November 1913, deals with the structure of polyatomic molecules, but it was really rather premature.

It was only natural that other physicists should not immediately recognize the significance of BOHR's research. But SOMMERFELD reacted positively: in September 1913 he wrote in a letter to BOHR that he was actually very sceptical of atomic models, but the calculation of the RYDBERG constant was a great achievement. H. G. J. MOSELEY expressed his results for the X-ray spectra of the elements with the aid of BOHR's theory. In December 1913 JEANS reported BOHR's 'convincing explanation of the spectral lines' to a meeting. He added, however, 'The justification of his theoretical assumptions is only the very ponderous one of success.' LORD RAYLEIGH (1842–1919) thought that people over sixty years of age should no longer participate in discussions of such matters. Otto STERN still recalls how in 1914 he and Max von LAUE promised each other that they would give up physics if there was 'anything in this nonsense of BOHR's'. PAULI was later to call this the '*Ütli-Schwur*' (the Ütli is a mountain near Zürich and '*Schwur*' means oath*). Happily, von LAUE and STERN did not keep their promise.

Other Possibilities

BOHR's ideas determined the direction of research in quantum theory for a whole decade. Taken together with quantum statistics as already developed they actually amounted to a provisional quantum mechanics of simple systems. So let us now go a little more carefully into the problem of chance and causality which we touched on in the introduction. We may regard as chance the fact that in the case of PLANCK, two things happened to coincide: he had thoroughly understood the concept of the entropy of equilibrium states and he lived in the very city in which precise measurements were carried out on black-body radiation. It was also chance that made BOHR work with RUTHERFORD. We may count as 'causal' the fact that the atom represented an important problem and that the black-body radiation law was very general, whereas for spectral series there were a number of useful but strange rules and a great deal of numerical data. Therefore, we

* *Translator's note:* The original '*Rütli-Schwur*' was taken by the Swiss in Schiller's *William Tell* (Act II, Scene (ii)).

shall ask the question that we posed in the introduction: what might have happened if...? How else could the subject have developed?

Possible starting points for the creation of quantum theory could have been: the series laws for spectra, the chemical and physical properties of atoms, the granular structure of short-wave light, the dependence of many properties on temperature, especially at low temperatures. Let us ignore the wave properties of matter for the time being as it was not until much later that these could be observed. We shall go into this in Chapter 11. Let us consider one by one these four possibilities.

Ever since RYDBERG, frequencies have been written as differences, and the general version (RITZ, 1908)

$$v = F(n_1...) - F(n_2...)$$

was implicitly already known before 1900, as we have seen. A comparison between the series of frequencies of a periodic motion and the observed spectral series could lead to the counterposition:

$$v = \tau v_1(E) \qquad v = F(n+\tau) - F(n)$$

as was finally effected by BOHR in 1913. The second expression of course tends to the form $v = \tau[F(n + 1) - F(n)]$ for large n. This counterposition could well be regarded as an indication of the enormous discrepancy between classical theory and reality. But it would perhaps also be used as a bridge. The formula $v_1 = dE/d\Phi$, with $\Phi = \oint p \, dx$, which holds for classical mechanics, was of course known only to a few people. But it might well have occurred to one of these few to compare the two expressions

$$v_1 = \frac{dE}{d\Phi} \qquad v_1 = \frac{\Delta F}{\Delta n} \ (\Delta n = 1)$$

which suggests the quantum theorem

$$\Phi \sim n \qquad E \sim F.$$

With the notation h for the missing constant, this would have amounted to:

$$\Phi = hn \qquad E = hF.$$

The actual frequencies would then have been asymptotically equal to the classical frequencies for large n. This would have been a quantum theory of the correspondence principle (just like

BOHR's) but without PLANCK or EINSTEIN, and omitting the harmonic oscillator, light quanta and specific heats. It would have been conceivable even without RUTHERFORD. It was possible to deduce $E \sim -1/n^2$ from the hydrogen series

$$\nu \sim \frac{1}{n_1^2} - \frac{1}{n_2^2}$$

and for large n

$$\nu_1 \approx \frac{dF}{dn} \sim \frac{1}{n^3} \sim |E|^{3/2}.$$

This would have led to the discovery of the well-known relationship between frequency and energy for the motion of a charged particle in a Coulomb field, and thus finally to the concept of the hydrogen atom in the RUTHERFORD form.

Just as the rigorous form of quantum mechanics developed logically from the BOHR correspondence principle (Chapter 10), so the whole of quantum mechanics could have been generated by the series laws for spectra.

The remaining properties of the atom, as we saw in the Survey (Chapter 1), give little direct help in the formulation of quantum theory. The chemical properties are certainly connected with spin and the shell structure. The stability of the atom could still not be understood in spite of RUTHERFORD's work, and the actual atomic radius a was a new incomprehensible unit. It was possible to understand the orders of magnitude of the ionization potential and the frequencies in terms of e, m and a. But without the additional help of the spectral laws, or of temperature dependence, it would probably have been impossible to make any further progress.

The situation was slightly more hopeful in the case of the light quantum. EINSTEIN conceived this idea as a result of an analysis of the WIEN radiation formula. Other results which clearly pointed to the existence of a light quantum had hardly been looked into at that time. The connection between the generating potential and the 'hardness' and ionizing action of X-rays was striking. But hardness could at first be related only qualitatively, or indirectly, to the frequency. However, as soon as it was possible to measure the wave lengths of X-rays (from 1912), it was possible to establish a relationship of the type

$$E_{kin} = h\nu.$$

The hypothesis that followed from this, that light could be absorbed or emitted only in quanta of energy hv, could lead to a derivation of PLANCK's radiation formula, as was performed by EINSTEIN in 1917. The transition to an atomic dynamics would then once again have required the spectroscopic combination principle.

In fact it was the quantum of action that was discovered, for black-body radiation. The very closely related phenomena that arise in the case of specific heats were far less informative, even after 1910, when it became possible to work with very low temperatures. The presence of the degrees of freedom for oscillation for a rigid body is actually slightly obscured by the fact that such a body has a large number of frequencies of vibration, with the result that it is not easy to spot a condition of the type $kT > hv$. When we introduce the degrees of freedom for rotation into the specific heats of biatomic gases it is certainly possible to observe a condition $IT > $ const. (h^2), but the rotator is not such a simple system for quantum theory as the harmonic oscillator. The deduction of the distribution of the states over the energy scale from the specific heats is really quite difficult, involving the deduction of an integrand from its integral. The NERNST heat theorem would have helped. It could have provoked the question of the entropy constant, as it did in reality. It was necessary to accept a natural unit of phase space. It was possible to calculate a unit per degree of freedom, rather inaccurately, from the empirical entropy of gases and call it h. And then the spectral laws would once again have been necessary to complete quantum mechanics.

Apart from black-body radiation only the spectral laws could have led from the empirical beginnings to the ultimate formulation of quantum theory.

[1] N. BOHR, Phil. Mag. **26**, 1, 476, 857 (1913). New edition of original essays, Gopenhagen 1963 with introduction by L. ROSENFELD

[2] cf. ROSENFELD's Introduction

[3] N. BOHR, Phil. Mag. **27**, 506 (1914)

[4] N. BOHR, Fysisk Tidsk. **12**, 97 (1914)

[5] H. G. MOSELEY, Phil. Mag. **26**, 1024 (1913), **27**, 703 (1914).

6. THE QUANTIZATION OF PERIODIC MOTION

Summary

BOHR's theory of the stationary states of atoms treated the combination principle for spectra

$$v = F(n_1 \ldots) - F(n_2 \ldots)$$

as a counterpart of the quantum relationship

$$h v = E(n_2 \ldots) - E(n_1 \ldots).$$

It also provided a method of calculating the energies $E(n)$ of the stationary states for those systems for which an expression $v(E)$ connecting the energy and frequency could be obtained from classical mechanics. In fact it required the classical frequencies $\tau v(E)$ and the quantum frequencies $[E(n + \tau) - (E(n)]/h$ to be asymptotically equal. This was achieved by deducting the values of $E(n)$ from the equation $h v(E) = dE(n)/dn$. This BOHR theory thus pointed the way for subsequent developments in quantum theory and introduced a well-defined period of its history, running from 1914 to around 1922. It is sometimes referred to as the 'Old Quantum Theory', the 'BOHR Theory', or sometimes the 'quantum mechanics of the correspondence principle'. This period was followed by what was called the 'crisis in Old Quantum Theory'.

The quantum theory of the correspondence principle was developed in a number of research centres. It was in Copenhagen that the major advances were made in the foundations of the subject, stimulated by EHRENFEST's adiabatic hypothesis. EHRENFEST had succeeded H. A. LORENTZ at Leyden in 1912. In Munich SOMMERFELD and his pupils applied the theory to atomic spectra and used this practical exercise as a basis from which to develop the theory.

At first a necessary condition for the applicability of the theory was the existence of the classical frequencies τv or, rather more generally, $\tau_1 v_1 + \tau_2 v_2 + \cdots$. This led to the development of a quantum theory that applied to systems executing periodic or multiply-periodic motion.

Simply-periodic Motion

The first thing that BOHR tackled was the hydrogen atom. It was indeed a system with three degrees of freedom. (It was possible to regard the nucleus as a fixed centre.) But it had a single fundamental frequency $v(E)$, analogous in this respect to a system with one degree of freedom. For systems of this kind there were the following formulations of a quantum theory. In statistics the region of phase space

$$\Delta \oint p \, dq = h \tag{1}$$

is equivalent to a single case. The states to be counted in this form of statistics were given by

$$\Phi = \oint p \, dq = 2 \oint E_{kin} \, dt = 2\overline{E_{kin}}/v = hn \tag{2}$$

and for large values of n this tended towards classical statistics. BOHR's formulation (with the classical frequency $v(E)$)

$$h v(E) = \frac{dE(n)}{dn} \tag{3}$$

immediately guaranteed the asymptotic agreement of the frequencies. It followed from the equation

$$dE = v \, d\Phi \tag{4}$$

which had already been used by HASENÖHRL, that (2) and (3) amounted to the same thing. BOHR discovered a more fundamental basis for the theorem $2\overline{E_{kin}} = hvn$ in the adiabatic invariance of Φ.

The adiabatic theorem in classical mechanics has a long history. It first appeared in the work of BOLTZMANN (1866) and CLAUSIUS (1871), actually in the thermodynamic distinction between internal and external variables, and in connection with the question of how the internal motion varied as the result of a change in an external parameter. In this connection the quantity

$$\Phi = 2\overline{E_{kin}}/v = \oint p \, dq$$

was significant. Here the following relationship held for an unchanged parameter and the transition to a neighbouring value of the energy:

$$\delta E = v \delta \Phi.$$

For slow 'adiabatic' change of the parameter, Φ remained constant throughout the actual motion. Thus for example RAYLEIGH was aware that E/v was constant for a pendulum whose length was gradually shortened.[1] This invariance emerged in a quantum context in 1911. EHRENFEST attempted to interpret the radiation formulae by means of a weight function in phase space (Chapter 2). He noted that the adiabatic change of a vacuum filled with radiation, as considered by WIEN, required the form $G(E/v)$ for this. EINSTEIN went on to answer a question posed by LORENTZ at the Solvay Congress in 1911 that for a slow change of parameter an oscillator state of energy hvn was again transformed into a state of energy hvn (see Appendix).

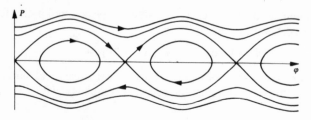

FIGURE 5:   THE (φ, P) GRAPH

In May 1913 EHRENFEST succeeded in justifying the expression $\overline{E_{\text{kin}}} = hvn/2$ or $P = h\dot{n}/2\pi$ for a free rotator by considering the adiabatic transition of an oscillation into a rotation by a decrease in the applied force. In November 1913 he posed the two general questions[2]: is there an adiabatic relationship for arbitrary periodic systems analogous to the invariance of E/v for the harmonic oscillator? Is it possible to preserve PLANCK's formula $E/v = hn$? An affirmative answer to the first question followed from the above theorem due to BOLTZMANN and CLAUSIUS, according to which

$$\oint E_{\text{kin}} \, dt = \overline{E_{\text{kin}}}/v$$

is adiabatically invariant. The answer to the second question was given by the formula

$$\overline{E_{\text{kin}}}/v = hn/2$$

as this holds for an harmonic oscillator, which can therefore be transformed adiabatically into a rotator. The ellipses E = constant in the (φ,P)-plane then become pairs of straight lines. The (φ,P) graph for a system undergoing an adiabatic transition is given in Figure 5. EHRENFEST thus changed his correct expression $P = (h/2\pi)n$ of May that year for the case of the free rotator, by assigning two circuits to the to-and-fro motion when it passed through a singularity. This gives $v = \dot{\varphi}/4\pi$ for the rotator, so that he put $E_{kin} = 2vP$ and thus obtained

$$P = hn/4\pi.$$

The pair of states $P = \pm(h/4\pi)n$ then together had the same weight as one oscillator state. It would have been correct to assign one circuit to the to-and-fro motion with the same weight and thus to write as before

$$P = hn/2\pi.$$

EINSTEIN was later to call EHRENFEST's principle the 'adiabatic hypothesis'.[3] The principle stated that it was possible to derive from the quantum states of one system the values of states of another system that could be reached by adiabatic transformation. EHRENFEST himself expressed it neatly in 1916[4]: the adiabatic change of a parameter, he said, transforms admissible paths into admissible paths. In this way it is possible to deduce from the quantization of a known system the quantization of some other system.

As early as 1914 BOHR had recognized in equation (4) the connection between his method (3) and the earlier formula (2) for phase extension.[5] In a paper written in 1916 but not published at the time he proposed to give a systematic representation of the quantum theory of periodic motion.[6] BOHR here made two equations of classical mechanics the basis of his approach:

$$\delta E = v\,\delta(2\overline{E_{kin}}/v) \tag{5}$$

for the theoretical transition to a neighbouring state of motion of the mechanical system, and

$$\delta(\overline{E_{kin}}/v) = 0 \tag{6}$$

for the actual transition under adiabatic change of a parameter. (6) enabled him to derive the equation

$$\Phi = \oint p\,dq = 2\overline{E_{kin}}/v = hn. \tag{7}$$

Asymptotic agreement between the classical and quantum-theoretical frequencies was guaranteed by (5):

$$\tau v = \tau \, dE/d\Phi \approx [E(n + \tau) - E(n)]/h.$$

In Copenhagen in 1916, it was felt that the quantum theory of periodic motion was firmly based on equation (7). BOHR's 1916 reflections were later incorporated in his important paper of 1918.[7]

As $\Phi = hn$ for the harmonic oscillator, it must hold for all systems that can be adiabatically transformed into one. As every value of n has the same weight for the oscillator, this must also hold for the other systems. The asymptotic validity of classical physics arises in statistics as a result of the equation

$$\Phi(n+1) - \Phi(n) = h$$

and for frequencies according to (5). The asymptotic transition to classical physics is also assumed to hold for intensities. The C_τ^2 of the Fourier series for the classical motion of one co-ordinate

$$x = \sum_\tau C_\tau \cos(2\pi \tau v t + \gamma_\tau)$$

approximately determine the intensities and thus the probabilities of the corresponding quantum transitions $(n + \tau, n)$. If for example in a classical motion, say for the harmonic oscillator, only $\tau = 1$ arises, the selection rule $\Delta n = \pm 1$ holds in quantum theory, and this may be expected to hold also for small values of n. The major portion of the 1918 paper then deals with systems with several fundamental frequencies. The expression 'correspondence principle' was subsequently used in two senses: in the narrower one as the method of determining intensities and as the guide for selection rules, and in the broader sense as a general method of establishing quantum equations in such a way that they tended asymptotically to the corresponding classical ones.

Difficulties

Quantization by means of the phase-integral (2) led to absurdities which might have suggested that it was not strictly true. It was not until later, when the comparison of prediction and experiment indicated the failure of the method, that these absurdities were actually recognized.

EHRENFEST and his colleagues demonstrated a number of cases in which the literal application of (2) led to ridiculous results.[8] In the example given by EHRENFEST and BREIT in 1922 for a free rotator, under suitable conditions the sense of rotation is reversed after every $f/2$ orbits. So apart from the reversal time T it has another period fT. Quantization by means of the phase-integral gives

$$\oint p\,\mathrm{d}q = 2\pi f P$$

and thus values of the angular momentum $P = hn/2\pi f$, while in the limit for large f only the values $hn/2\pi$ can arise. The correspondence principle, however, gives a perfectly sensible explanation. As the motion with frequency fv has a particularly large amplitude in the Fourier series the transitions $\Delta n = \pm f$ receive a particularly large probability as a result of the change

$$P = h/2\pi$$

in the angular momentum. In the version given by EHRENFEST and R. C. TOLMAN the states for which n is not a multiple of f have 'weak quantization', i.e., a low statistical weight, which tends to zero if f becomes large. We should nowadays point to the connection between this difficulty and the whole question of the idealization of the kinematic model. Rotators of atomic dimensions simply do not have a physical representation of this kind. A similar situation arises in the case of a torsional vibration as considered by EHRENFEST and TOLMAN, where a rotator makes a large number of circuits before reversing its sense of rotation. EHRENFEST and TOLMAN went on to consider a free rotator of such symmetry that it displays the same configuration after a rotation through $2\pi/\sigma$. Strictly speaking, the correspondence principle requires that $\Delta n = \pm \sigma$, with the result that it is possible to deduce that only the states $n = 0, \sigma, 2\sigma,\ldots$ can arise. In the limiting transition from approximate σ-symmetry to complete symmetry EHRENFEST and TOLMAN assumed a quantization that again became weaker. Today we know that the restriction to $n = 0, \sigma, 2\sigma,\ldots$ is a consequence of the BOSE statistics of indistinguishable particles and that the limiting transition from distinguishable to indistinguishable particles can therefore not occur. Other examples due to EHRENFEST and TOLMAN were concerned with the disturbance of periodic motion as a result of random external influences, which led to inexactness in the energy.

Gradually BOHR came to feel that the phase-integral should not be taken anything like so literally, and in Göttingen this feeling was expressed in the form of a slogan: 'up with the correspondence principle, down with the phase integral!'

Real absurdities would have resulted from an examination of even more unusual cases of the potential energy $V(x)$, for example a potential barrier (see Appendix). EHRENFEST had in fact already suggested in 1913 that it might be necessary to modify the adiabatic law for such steps, but no discussion of the quantum theory of such systems can be proved to have taken place before 1926.

Multiply-periodic Motion

BOHR related the atom with a single electron to the classical formula $v \sim |E|^{3/2}$. This hardly satisfied his contemporaries. They did not immediately see the connection with the method with which they were much more familiar: $\oint p \, dq = hn$. And so it was that a process was set in motion which was certainly inspired by BOHR, but which in fact had more to do with PLANCK.[10]

In the spring of 1915 William WILSON put

$$\oint p_k \, dq_k = h n_k \tag{8}$$

for every pair of conjugate variables, in order to establish a common theoretical basis for the BOHR and PLANCK theories for a system with several degrees of freedom. He did not, however, test its applicability any more closely, nor did he carry out any calculations of atomic states. Nobody noticed a similar piece of work produced at the same time by J. ISHIWARA. In the winter of 1915–6 PLANCK and SOMMERFELD, unaware of WILSON's result, tackled systems with several degrees of freedom. SOMMERFELD assumed (8) to hold in general and used it to deal with motion in a central field. With plane polar co-ordinates r, φ to which the variables p_r and angular momentum P are conjugate 'momenta', and thus putting

$$\oint p_r \, dr = h n_r \qquad \oint P \, d\varphi = 2\pi P = h n_\varphi \tag{9}$$

he obtained the same energy values for a Coulomb field as BOHR had obtained:

$$E = 2\pi^2 m Z^2 e^4 / h^2 n^2 \qquad n = n_r + n_\varphi.$$

Equal values of n corresponded to equal lengths of the major axes of the elliptic paths. n_φ distinguished between them with respect to eccentricity, which thus took discrete values. The treatment using three-dimensional polar co-ordinates r, ϑ, φ and three phase integrals yielded the same expression for the energy, but with

$$n = n_r + n_\vartheta + n_\varphi$$
$$2\pi P = h(n_\vartheta + n_\varphi) \qquad 2\pi P_z = h n_\varphi$$

where P_z was the angular momentum about the axis of the azimuthal angle φ. In common with P_z/P the inclination of the orbital plane could assume only discrete values. This suggested the idea of 'direction quantization'. Of course SOMMERFELD was very unhappy about the BOHR argument using $v \sim |E|^{3/2}$. His own work allowed him to regard the atom with a single electron as partially understood.

He was aware that the BALMER series for hydrogen did not consist of single lines. But it was only the use of relativistic mechanics with its dependence of mass on velocity that was able to explain this 'fine structure' using the quantum theorems (9). E now depended on n_r and n_φ. The predictions agreed very closely with the experimental evidence even in the case of the ionized helium atom. It was a not insignificant factor in the history of quantum theory that SOMMERFELD's calculations without electron spin and with a provisional quantum theorem gave largely the same results as the rigorous calculations were later to do. SOMMERFELD's work and the result $E(n_r, n_\varphi)$ also gave an indication of how to interpret those spectra which were unlike hydrogen, where the deviation from the Coulomb field was a result not of the relativistic correction but of the influence of the other electrons.

Further confirmation of the predictions and a methodological advance came with the researches of Karl SCHWARZSCHILD and of P. S. EPSTEIN which date from the spring of 1916.[11] They applied to the classical motion the methods of mathematical astronomy, celestial mechanics, which had been developed around 1870. SCHWARZSCHILD had already noted in 1914 that a hydrogen atom in a homogeneous electrical field was the limiting case of a system of a single particle in a field with two centres of force, proportional to $1/r^2$, i.e., of a system which had been dealt with in celestial mechanics by means of elliptical co-ordinates. In 1916 SCHWARZS-CHILD and EPSTEIN treated the hydrogen atom in an electric field by applying formula (8) for parabolic co-ordinates (the limiting case of elliptic co-ordinates). Their method amounted to transforming the variables q_k, p_k into new variables w_k, I_k for a system with Hamiltonian $H(p_1, p_2, ..., q_1, q_2, ...)$ by a 'canonical transformation' in such a way that the Hamiltonian depended only on the I_k. According to the canonical equations of motion the I_k are then constant throughout the motion and the w_k are linear functions of time. For oscillatory or orbital motion of the w_k these can be chosen in such a way that they increase by 1 in a period, so that the associated frequency is

$$v_k = \dot{w}_k = \frac{\partial E(I_1, I_2 ...)}{\partial I_k}.$$

These variables are called angle and action variables. For one degree of freedom I is the quantity which we earlier denoted by Φ:

$$I = \oint p \, \mathrm{d}q.$$

For several degrees of freedom, if the procedure can in fact be carried out, we have

$$I_k = \oint p_k \, \mathrm{d}q_k.$$

Thus it was that SCHWARZSCHILD and EPSTEIN were able to put

$$I_k = h n_k.$$

The results of their calculations fitted in well with STARK's measurements for the STARK effect for hydrogen.

In his major paper of 1918 BOHR considered 'conditionally periodic systems', which were also called multiply-periodic systems.[7] These include those systems that consist of independently oscillating sub-systems, for which

$$H = \sum_k H_k(p_k, q_k)$$

as for example the anisotropic oscillator with $V = V_x(x) + V_y(y)$. BOHR expected

$$I_k = \oint p_k \, \mathrm{d}q_k = h n_k \qquad (10)$$

to hold in these cases.

More generally these include systems that admit of a 'separation' into co-ordinates q_k, as occurs for example in the case of a particle in a central field. Each p_k then depends only on its q_k and on the constants of the motion. If the system remains finite the q_k oscillate or orbit, which would seem to suggest (10). This is supported by the fact that these I_k are adiabatically invariant, as shown by BURGERS when he took up some work of EHRENFEST in 1917. If the motion completely fills a region of co-ordinate space —as in the case of the anisotropic oscillator, and for central motion if we restrict ourselves to a plane (Figures 6 and 7)—the I_k are uniquely determined. In the case of 'degenerate' systems, which have fewer fundamental frequencies than degrees of freedom— such as the isotropic oscillator, motion in a Coulomb field, and general central motion in three dimensions—the separation variables may be chosen in different ways and the transition from one choice to another for a given motion changes the values of the I_k. BOHR regarded every motion as stationary if it had the property that for some choice of the separation variables the I_k were integral multiples of h. If an additional field is gradually

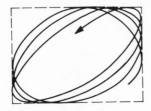

FIGURE 6: THE PATH FOR AN ANISOTROPIC OSCILLATOR

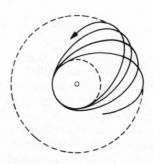

FIGURE 7: THE PATH FOR A CENTRAL FIELD

imposed in such a way as to preserve the degeneracy a 'stationary' motion should then develop, related to the one that has integral values of the I_k/h for the (now uniquely) chosen co-ordinates. This would not contradict the adiabatic law, because, as a result of the degeneracy, there are always degrees of freedom to which an even slower motion corresponds if we impose the field more slowly. This observation is characteristic of BOHR. It shows how keenly he sought to preserve the validity of classical mechanics for stationary states. BOHR concluded that as the states with particular values of n_1, n_2,... could be adiabatically transformed into states for which the degrees of freedom were completely independent or into states with other values of n_1, n_2,..., then all non-degenerate states had equal statistical weight. The weight of degenerate states was determined by the number of non-degenerate states into which they split as a result of the approach of a suitable weak additional field. For large quantum numbers this agrees with classical statistics, where the number of cases is given by $\delta W = \delta I_1 \delta I_2 \ldots$.

For large values of n_1, n_2,... the quantum frequencies

$$\nu = [E(n_1 + \tau_1, n_2 + \tau_2 \ldots) - E(n_1, n_2 \ldots)]/h \qquad (11)$$

tend to the classical frequencies

$$\nu = \Sigma \tau_k \nu_k \qquad (12)$$

because the classical ν_k are determined by the formula

$$\delta E = \Sigma \nu_k \delta I_k \qquad (13)$$

and thus we have, by (10),

$$\nu = \frac{1}{h} \left[\tau_1 \frac{\partial E}{\partial n_1} + \tau_2 \frac{\partial E}{\partial n_2} + \cdots \right].$$

These are thus also according to (11) the quantum-theoretical frequencies for large n_k. BOHR also noted the agreement between his results and those of SCHWARZSCHILD and EPSTEIN.

BOHR went on to assume for the intensities, and thus for the probabilities of spontaneous transitions (the expression was actually used) that they tended towards the classical values for large n_k. These classical values were given by the values of $C_{\tau_1 \tau_2} \ldots$ in the fourier series for the electric moment

$$x = \Sigma C_{\tau_1 \tau_2} \ldots e^{2\pi i (\tau_1 \nu_1 + \tau_2 \nu_2 + \ldots) t}.$$

In particular the corresponding transitions should not occur even for small values of $n_1, n_2, \ldots$ if certain of the $C_{\tau_1 \tau_2} \ldots$ were zero.

If the system is invariant with respect to rotation about a fixed axis (the z-axis) the component of the angular momentum is a constant of the motion and $2\pi P_z$ is an action variable. In quantum theory, we thus take

$$P_z = (h/2\pi)m \qquad m = 0, \pm 1, \pm 2, \ldots$$

Conjugate to the canonical variable P_z is an angle that must increase monotonically, by symmetry. In the Fourier series for the z-co-ordinates or the z-component of the electric moment the orbital frequency does not occur. For dipole radiation with oscillation parallel to z the selection rule $\Delta m = 0$ thus holds. In the Fourier series for the component of the moment perpendicular to z the only terms that arise are $\tau = \pm 1$. And thus the selection rule $\Delta m = \pm 1$ follows for dipole radiation with oscillation perpendicular to z.

The Position around 1922

It was characteristic of current quantum theory that it was capable of dealing only with systems that had periodic or multiply-periodic motion. The 'stationary states' were selected from the possible classical motions by 'quantum conditions' $I_k = hn_k$. If the classical motion could not be calculated rigorously, it was approximated by a 'perturbation calculation' in the neighbourhood of a calculable motion. This perturbation method was based on that of astronomy and the angle and action variables of the simpler 'undisturbed' system were mostly used. Theoretical physicists were busily studying the books by Henri POINCARÉ and by C. L. CHARLIER on the mathematical methods of astronomy. A method of perturbation calculation for quantum theory had been developed by BOHR in his great paper of 1918. BORN and his collaborators worked on it very systematically in 1921–2, as did P. S. EPSTEIN in 1921. They developed the Hamiltonian in the neighbourhood of an 'undisturbed system',

$$H = H^{(0)} + \lambda H^{(1)} + \lambda^2 H^{(2)} + \ldots$$

and carried out the individual steps at each level of approximation. EPSTEIN applied perturbation methods to the case of a periodic external disturbance, the theory of optical dispersion.

He immediately realized the difficulty, which was that the calculation gave the frequencies that were generated by the dispersion instead of the classical frequencies of the stationary orbits, while in fact these ought to be the frequencies of emission and absorption.

The source of the difficulties and absurdities that gradually emerged lay of course in the assumption that the stationary states were particular classical states. So it was necessary to accept certain restrictions in the choice of the n_k when the classical motion involved something physically impossible. The degenerate states displayed certain difficulties, as they had fewer fundamental frequencies than degrees of freedom. However we choose the separation variables we are led to equal energy values but to different stationary states. SOMMERFELD tried to give even the numbers n_k, which did not enter into the energy, a physical meaning by regarding a degenerate system as the limiting case of a non-degenerate one. As late as 1924 he was referring to the discrete values of the eccentricity of the elliptical orbit for the hydrogen atom. BOHR on the other hand attributed no physical significance to information that was obtained as a result of the arbitrary choice of co-ordinates. The absurdities, which included the difficulty indicated by EPSTEIN's dispersion theory, were the result of the fact that the stationary orbits with classical frequencies $v(E,...)$ emitted or received radiation involving the quantum frequencies $v(n + \tau, n)$. The effect of one atomic system on another ought then in reality to lead to the quantum frequencies, while in the perturbation methods that were then in vogue the classical frequencies were used. These inconsistencies were later to become the source of an improvement in the theory (Chapter 10). The absurdities that arose in the case of potential barriers were still not taken into account by Wolfgang PAULI and K. F. NIESSEN in their attempt to calculate states of the hydrogen ion molecule (H_2^+) in 1921, a system of two nuclei and one electron.[13]

[1] LORD RAYLEIGH, Phil. Mag. **3**, 338 (1902)

[2] P. EHRENFEST, Verh. D. Phys. Ges. **15**, 451 (1913), Proc. Amst. **16**, 591 (1913)

[3] A. EINSTEIN, Verh. D. Phys. Ges. **16**, 820 (1914)

[4] P. EHRENFEST, Proc. Amst. **19**, 576 (1916), Ann. d. Phys. **51**, 327 (1916), Phil. Mag. **33**, 500 (1917)

[5] N. BOHR, Phil. Mag. **27**, 506 (1914)

[6] Abhandlungen über Atombau, p. 123

[7] N. BOHR, Dansk. Vid. Selsk. Skr., natw. o. math. Afd. 8. R. IV 1, 1918, part I reprinted in Sources edited by B. L. van der WAERDEN

The Quantization of Periodic Motion

[8] P. EHRENFEST and G. BREIT, Proc. Amst. **23**, 989 (1922), Z. Phys. **9**, 207 (1922)
P. EHRENFEST and R. C. TOLMAN, Phys. Rev. **24**, 287 (1924)
[9] P. EHRENFEST, Proc. Amst. **16**, 591 (1913)
[10] W. WILSON, Phil. Mag. **29**, 795 (1915)
J. ISHIWARA, Tokyo Sug. But. Kizi **8**, 106 (1915)
A. SOMMERFELD, Sitz. Ber. München **1915**, 425, 457, Ann. d. Phys. **51**, 1, 125 (1916)
M. PLANCK, Verh. D. Phys. Ges. **17**, 407, 438 (1915), Ann. d. Phys. **50**, 385 (1916)
W. PAULI, Naturwiss, **13**, 129 (1948)
[11] K. SCHWARZSCHILD, Sitz. Ber., Berlin **1916**, 548, Verh. D. Phys. Ges. **16**, 20 (1914)
P. EPSTEIN, Ann. d. Phys. **50**, 489 (1916), **51**, 168 (1916), Verh. D. Phys. Ges. **19**, 116 (1917)
[12] M. BORN and E. BRODY, Z. Phys. **6**, 140 (1921)
P. S. EPSTEIN, Z. Phys. **8**, 140 (1921)
M. BORN and W. PAULI, Z. Phys. **10**, 137 (1922)
[13] W. PAULI, Ann. d. Phys. **68**, 177 (1922)
K. F. NIESSEN, Diss. Utrecht 1922.

7. THE INTERPRETATION OF
SIMPLE SPECTRA

Atomic Structure and Spectral Lines

'ATOMIC Structure and Spectral Lines' was the title of SOMMERFELD's book. The publication of the first edition (1919), the third (1922) and the fourth (1924) each marks an important stage in the development of quantum theory. They belong to the period of the old quantum theory, which covers roughly the decade following BOHR's first paper on the structure of the atom. The prediction and evaluation of atomic properties, in particular that of the spectral terms, in an approximation that treats each electron separately in a static field of force belong to this period. The semi-empirical interpretations of the general features of atomic spectra, of chemical properties, and thus of the periodic system, also belong to this period. During this period new facts were discovered which made it possible to understand, for example, the shell structure and electron spin.

On the experimental side there was chemistry, in particular the periodic system and the theory of valence. Atomic spectra had been cleared up in the years 1885 to 1895 by BALMER, RYDBERG, KAYSER and RUNGE, PASCHEN, Alfred FOWLER, SAUNDERS and others, for those cases where the lines lay in the visible and near ultra-violet regions. PASCHEN (1900–10) dealt in particular with the infra-red spectra. Those of the higher ultra-violet region were investigated by LYMAN between 1910 and 1920. This extension of the accessible region of frequencies rounded off the knowledge of spectral series. The investigation of X-ray spectra was carried out mainly by MOSELEY, BARKLA and SIEGBAHN between 1913 and 1922.

In the decade which followed BOHR's work, quantum theory was able to explain the terms of the simple line spectra as the consequence of the motion of an external electron in a central field. Two quantum numbers, which are nowadays called n and l, corresponded to the two fundamental frequencies of this motion. The theory interpreted more refined features—the 'fine structure'

—as the effect of the internal electrons of the atomic core* and attempted to make do with three quantum numbers, n, l, j. X-ray spectra could also be disposed of in this way. Towards the end of this period there emerged increasingly complicated properties of the spectra. Nowadays we realize that three very different features of atomic physics were jumbled up together. These were the spin of the electron, the PAULI exclusion principle and to some extent the axiomatic foundations of quantum theory. The fact that these three factors occurred in the way they did, inextricably interwoven, made it difficult to give a satisfactory explanation of the problems that arose.

Quantum theory was now in a position to be tested out in the field of band spectra, initially of biatomic molecules (HEURLINGER, LENZ and KRATZER).

The n, l *Representation*

The view that simple spectra are fundamentally related to the motion of a single electron, the 'series electron' or 'emitting electron', belongs here, as do the introduction of two quantum numbers, the selection rule $\Delta l = \pm 1$ and the assignment of l-values to the s, p and d series and finally the estimation or evaluation of the terms.

The RYDBERG form $R/(n + \alpha)^2$ for the spectral terms, with a value of R that was almost constant for all elements, had already been explained by BOHR. He said that the series spectra were related to the motion of an external electron, which involved roughly the same force field as in the hydrogen atom.

It was SOMMERFELD who, in 1915, introduced two quantum numbers by idealizing the motion of an electron as motion in a central field and by using quantum conditions of the form

$$\oint p_r \, dr = h n_r \qquad \oint p_\varphi \, d\varphi = 2\pi P = h n_\varphi$$

where p_r and $p_\varphi = P$ are the variables conjugate to the plane polar co-ordinates r, φ. He let them take the values $n_r = 0, 1, 2, \ldots$; $n_\varphi = 1, 2, 3, \ldots$; $n_\varphi = 0$ would, of course, have given a straight line passing through the nucleus. He obtained the same expression for the energy as BOHR, $E \sim -1/n^2$, where $n = n_r + n_\varphi$.

* *Translator's note:* The term 'core' is used to denote the nucleus of the atom together with any complete, i.e. stable, shells.

At roughly the same time (1918) A. RUBINOWICZ in the SOMMERFELD school and BOHR gave a selection rule for the angular momentum quantum number, which, following BOHR, was mostly called k and later l.[1] RUBINOWICZ showed that an electromagnetic spherical wave can only transfer an angular momentum of 0 or $\pm h/2\pi$ when it receives or gives up an energy $h\nu$. He deduced from this that the angular momentum quantum number could only vary as between 0 and ± 1. In the spirit of his correspondence principle, although it was not yet known as such, BOHR concluded that the path of an electron in a central field was a plane 'rosette' and that the motion had the frequency

$$\nu = \tau\nu_r \pm \nu_\varphi.$$

Because of the uniform orbiting of the pericentre the frequency of φ is related only to the factor ± 1. This gives the selection rule

$$\Delta l = \pm 1.$$

In his book (1st edition, 1919) SOMMERFELD compared the two points of view and acknowledged BOHR's version as superior.

In 1915 SOMMERFELD had already essentially given the method of assigning angular momentum quantum numbers $l = 1, 2, 3...$ to the s, p and d terms in spectroscopy. The series for simple spectra

$$
\begin{array}{llll}
2p & 3p & 4p & 5p \quad ... \\
 & 3d & 4d & 5d \quad ... \\
 & & 4f & 5f \quad ...
\end{array}
$$

fitted well with the inequality $l \leq n$, which was theoretically required if one took $l = 2, 3, 4$ for the p, d and f terms. It was at the very latest after the selection rule $\Delta l = \pm 1$ was known that the value $l = 1$ for the s term became obvious. The qualitative interpretation of simple series spectra was completely satisfactory by about 1918.

SOMMERFELD tried in 1915 to estimate the values of the terms by means of a series

$$V(r) = -\frac{e^2}{r} - \frac{A}{r^3} - \frac{B}{r^5}$$

for the potential energy.[2] He was able to obtain the form for the terms given by RITZ in 1903, written in the form of an energy as

$$E(n, l) = -\frac{Rch}{[n - a(l) - b(l)E]^2}$$

where $a(l)$ was determined by A, $b(l)$ by A and B. None of the attempts to calculate the function $V(r)$ from empirical values of a and b was entirely successful. In general a was actually not a small quantity. SCHROEDINGER in 1921 realized that the essential point for the interpretation of the large 'RYDBERG corrections' was that the s orbits dipped deep into the atom.[3] A model with a point charge in the centre and a concentric spherical shell with electric charge $-8e$ in which the electron both outside and inside traversed an arc of an ellipse (Figure 8) gave for the s series of Na: $n^* = n - a = 1, 26, 2, 26...$, i.e., a quantitative disagreement. He had also not even assumed the correct value for n.

The correct values for n, $n = 2, 3, 4...$ for the ground states of Li, Na, K,... were given by BOHR in 1921 (he had not obtained them in 1920, nor had LANDÉ).[4] As the 'effective quantum number' n^* defined by the equation $E = -Rch/n^{*2}$ increases along the series H, Li, Na, K, Rb, Cs, although the field in the interior of the atom must strengthen he decided on $n = 1, 2, 3, 4, 5, 6$ for the six named elements. Rough estimates of the 'RYDBERG corrections' $n - n^*$ supported this method. Thus by 1921 the n, l model of simple spectra was understood in order-of-magnitude terms.

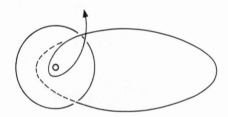

FIGURE 8:   THE s ORBIT

Predictions by Erwin FUES in 1922 showed that a potential $V(r)$ could be found for both Na and Mg^+ which brought the calculated terms in BOHR's enumeration of n into agreement with the measured values.[5] This meant that the large RYDBERG corrections were understood for the s and p terms. In 1924 M. BORN and W. HEISENBERG were able to interpret the slight corrections of the f terms and (to some extent) of the d terms as a consequence of the polarization of the core by the external electron.[6]

SOMMERFELD had already given one interpretation of the RITZ formula, and thus of the relationship

$$n - n^* = a + bE$$

by means of his series expansion of the potential in a central field. But the formula held more generally than did this series. It was in 1922 at the latest that BOHR gave an explanation which is both of intrinsic interest and typical of his application of the correspondence principle.[7] He likened the external elliptic arc which lies in the field of the nuclear charge e and which can be extended to a whole ellipse to the actual path. The number n^* corresponds to the energy of the external ellipse, and the frequency v^* is:

$$v^* = \frac{dE}{h\,dn^*} \qquad \frac{1}{v^*} = \frac{h\,dn^*}{dE}$$

while for the actual path we have

$$v = \frac{dE}{h\,dn} \qquad \frac{1}{v} = \frac{h\,dn}{dE}.$$

The difference between the periodic times of the two paths depends only slightly on E. We can expand it in terms of E and obtain:

$$\frac{1}{hv} = \frac{1}{hv^*} + b + cE + \cdots$$

$$\frac{dn}{dE} = \frac{dn^*}{dE} + b + cE + \cdots$$

$$n = n^* + a + bE + cE^2 + \cdots$$

The weakness in this argument clearly lies in the fact that the important quantity a is not interpreted. Its strength lies in the fact that the RYDBERG principle is explained for molecules as well, without any need to assume a central field. WENTZEL obtained a related result in 1923 using the phase integral for the central field.[8] For the external ellipse we have

$$(n^* - l)h = \oint p_r^* \, dr$$

where p_r^* is given by $V = -e^2/r$, and for the actual path

$$(n - l)h = \oint p_r \, dr$$

where p_r is given by the actual value of V. Once again, the difference between the two expressions hardly depends on E and it can be expanded as:

$$n - n^* = a + bE + \cdots$$

This was confirmed by later proofs using the SCHROEDINGER equation and the WKB approximation (Chapter 12).

n, l, j *Representation of Optical Spectra*

Nowadays we use the n, l, j method in deriving terms from the motion of single electrons in the field of the atomic core, assumed spherically symmetrical. The values $j = l \pm 1/2$ come from the spin. Though at that time spin was not yet known, it was recognized that many terms were double or triple.

For doublets and triplets there was RYDBERG's law of 1897: elements with odd valence numbers have doublet series, while elements with even valences have triplet series. The extension of the measurements into the ultra-violet region once again made the rule uncertain. The spectra of Mg and Ca contained single terms, doublet terms and triplet terms. But in 1914 A. FOWLER, inspired by BOHR's theory of the He^+ lines, showed that for the doublet series of these elements RYDBERG formulae held with $4R$ and that they therefore belonged to Mg^+ and Ca^+.[9] The qualitative aspects of spectra were eventually combined in the KOSSEL-SOMMERFELD laws of displacement and interchange[10]: the spectrum of a simply ionized atom has the same form as the spectrum of the neutral atom with an equal number of electrons that precedes it in the periodic system. Odd numbers of electrons lead to doublet spectra, even numbers give singlet and triplet spectra. Thus Na and Mg^+ have terms s, p_1, p_2, d_1, d_2.... Mg and Al^+ have one triplet system s, $p_1, p_2, p_3, d_1, d_2, d_3$,... and one singlet system s, p, d,.... s terms are invariably simple. Helium did not fit into this scheme as it was still necessary to treat the spectrum of orthohelium as a doublet. It later transpired that two terms of a triplet happened to lie very close together.

How would it be possible to interpret this 'fine structure' of the terms? It was certain that it indicated a deviation from pure central motion of an electron, in other words the atomic core could not be spherically symmetrical. Alfred LANDÉ (1919)

attributed an angular momentum to the atomic core. D. ROZH-
DESTVENSKY invested it with a magnetic field in 1920. In 1920 BOHR
spoke more generally of the existence of a third frequency as a
result of the perturbation of an otherwise plane electron path by

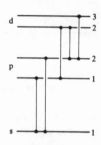

FIGURE 9:   COMBINATIONS IN THE DOUBLET SPECTRUM

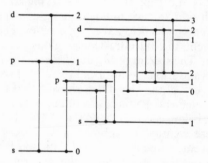

FIGURE 10:   COMBINATIONS IN THE SINGLET–TRIPLET SPECTRUM

interaction with the remaining electrons.[11] In 1920 SOMMERFELD
discovered a formal representation by introducing an 'internal
quantum number' (which we shall write without further ado as j)
with the selection rule $\Delta j = \pm 1$, 0.[12] The pairing of values of j
with the values of l was carried out with the aid of the 'forbidden'
combinations. Figure 9 shows the permitted combinations for a
doublet system. A p triplet combined differently with the simple
s term of the triplet system and the s term of the singlet system.
LANDÉ managed to explain the possibilities (Figure 10) using the
additional prohibition of the combination $0 \rightarrow 0$ of the j values.[13]

The free additive constant in j was chosen differently by different authors, so that for example according to BOHR $1 \rightarrow 1$ was the forbidden combination.

LANDÉ (1921) attributed the interior quantum number to the total angular momentum of the atom, which was composed of the angular momentum of the path and an angular momentum of the atomic core. BOHR too was led to this interpretation.[14]

By 1921 the structure of the terms of simple spectra was understood by modelling them as the combination of the angular momenta of the path and of the atomic core. There were not only the simple spectra of the atoms in the first three columns of the periodic table but also the more complicated spectra. One of the first to be resolved into terms was that of Ne. It was still not possible to explain the values of j.

n, l, j *Representation of X-ray Spectra*

When electrons or X-rays collide with matter, we observe sharply defined frequencies peculiar to the elements involved, as well as a continuous spectrum. In 1913 MOSELEY discovered that these frequencies behaved monotonically as the atomic number Z, in the case of the 'hardest' line, the first line of the 'K-group'

$$\nu = \tfrac{3}{4}Rc(Z-s)^2 \quad (s=1).$$

He recognized the agreement with BOHR's theory. A little later he was able to show the corresponding result for the L-group.[15] The occurrence of absorption fringes instead of absorption lines was explained by W. KOSSEL in 1914 and 1916, as the consequence of the fact that in absorption an internal electron is separated from the electron, while for emission electrons recede to positions far within the atom.[16] SOMMERFELD applied his theory, which took account of the relativistic correction, to the X-ray term and verified the predicted behaviour of the splitting with Z. He was, however, unable to explain the whole question of term multiplicity.[17]

Finally, in 1921, SMEKAL, COSTER and in particular WENTZEL solved all the X-ray terms analogously to the alkali terms.[18] They were able to draw on advances in experimental research and on the theoretical groundwork of KOSSEL and SOMMERFELD. Table 2 gives the method of assigning quantum numbers n, l and j to the X-ray terms K, K_{322}, M_{54321}. Adjacent terms whose difference corresponds to SOMMERFELD's theory ('regular' or 'relativistic' doublets) are joined by a ∩. Other adjacent terms are linked by a ∪ ('irregular' doublets). If we imagine the distance between the irregular doublets as tending to zero we obtain the multiplicity as given by SOMMERFELD's theory. The combination rule $\Delta j = \pm 1, 0$ holds for the quantum number j (as we shall call it) which determines the difference between the terms. If we imagine the relativistic doublets as tending towards each other we obtain the central field situation. We then have $\Delta l = \pm 1$ for the relevant quantum number.

TABLE 2: X-RAY TERMS

	K	L	M	N	
	1	3 2 1	5 4 3 2 1	7 6 5 4 3 2 1	
n	1	2	3	4	
l	1	1 2 2	1 2 2 3 3	1 2 2 3 3 4 4	$\Delta l = \pm 1$
j	1	1 1 2	1 1 2 2 3	1 1 2 2 3 3 4	$\Delta j = \pm 1,0$

Nowadays we count l as being one less ($l = 0,1,2\ldots$ for s, p, d,....) and j as being less by $\frac{1}{2}$ ($j = l \pm 1/2$). With the aid of the n, l, j scheme it was possible to understand the multiplicity of the terms in the optical series spectra for atoms with one, two or three external electrons, as well as the multiplicity of the terms for X-ray spectra.

The X-ray terms appeared to be the s, p, d... terms for a single electron with an additional relativistic splitting. It was as yet impossible to understand this, as in SOMMERFELD's theory it was caused by the different amounts by which the s, p, d... orbits dipped (for different values of l). It is now believed to depend on a different orientation with respect to the core of the atom (i.e., different values of j).

The splitting of optical doublets, triplets and relativistic X-ray doublets increases rapidly with the atomic number. In the limiting case of nuclei with small mass the spectra could thus be explained by assuming a central motion of the emitting electron.

It is impossible to assess the extent to which this partial understanding of optical spectra and X-rays satisfied any of the leading workers in the field in 1921–2. The third edition of SOMMERFELD's book, completed at the beginning of 1922, written in optimistic vein, gave an exhaustive presentation of series spectra, the anomalous ZEEMAN effect and X-rays, but regarded his presentation as 'basically biased towards the quantum-theoretical point of view, but essentially empirical'. The anomalous ZEEMAN effect had a particularly stimulating effect after the discovery of the higher multiplets. Chapter 9 will deal with this. But we shall first look into a number of other grounds for optimism (Chapter 8).

More Experiments

The FRANCK and HERTZ electron collision experiment in 1914 showed that a measured kinetic energy was connected with a spectral frequency by the equation $\Delta E = hv$ (Chapter 3). BOHR immediately saw the significance of this.[19] It was proof positive of the fact that the spectral terms actually corresponded to the energies. Most conveniently, in the years that followed, measurements of electron collisions were followed by spectroscopic measurements, so that, for example, it was shown that the lowest state of the orthohelium spectrum lay well above the ground state of the helium atom.

Another very important piece of work was a proof of the directional quantization of a silver atom in a magnetic field by Otto STERN and Walter GERLACH.[20] According to classical theory the magnetic moment m generated by an orbiting electron was some definite fraction of its angular momentum P. Both of course depended on the area

$$m = evF \qquad P = \mu \cdot 2vF$$

(μ is the mass of the electron). The equation thus read

$$m = \frac{e}{2\mu} P.$$

In quantum theory P was an integral multiple of $h/2\pi$. Orbits with a single quantum as were assumed in the cases of the ground states of H, of the alkalis Cu, Ag, Au, should therefore display a magnetic moment of size

$$m = \frac{eh}{4\pi\mu}$$

(the Bohr magneton) and align themselves parallel to the component $\pm eh/4\pi\mu$ of an external magnetic field. It was not at that time considered necessary to exclude the case with component 0. Stern suggested in 1921 that this 'directional quantization' could be proved experimentally by allowing atoms to pass through an inhomogeneous magnetic field. Stern and Gerlach carried out the experiment using silver atoms and in 1921 were able to demonstrate the two alignments. In 1922 they proved that the magnetic moment was indeed of size $eh/4\pi\mu$. They saw this as the verification of the angular momentum $h/2\pi$ ($l = 1$) of the ground state of the silver atom. It was later found necessary to modify this explanation somewhat, in the light of the anomalous Zeeman effect.

[1] A. Rubinowicz, Phys. Z. **19**, 441, 465 (1918)
    N. Bohr in the 1918 paper
[2] A. Sommerfeld, Sitz.-Ber. München **1915**, 457
[3] E. Schroedinger, Z. Phys. **4**, 347 (1921)
[4] N. Bohr, Z. Phys. **9**, 1 (1922)
[5] E. Fues, Z. Phys. **12**, 1 (1922), **13**, 211 (1923)
[6] M. Born and W. Heisenberg, Z. Phys. **23**, 388 (1924)
[7] reproduced in Born, Atomic Physics
[8] G. Wentzel, Z. Phys. **19**, 53 (1923)
[9] A. Fowler, Phil. Trans. Roy. Soc. A **214**, 225 (1914)
[10] W. Kossel and A. Sommerfeld, Verh. D. Phys. Ges. **21**, 240 (1919)
[11] A. Landé, Verh. D. Phys. Ges. **21**, 585 (1919)
     D. Roschdestwensky, Verh. Opt. Inst. Petrograd **1**, (1920)
     N. Bohr, Z. Phys. **2**, 423 (1920)
[12] A. Sommerfeld, Ann. d. Phys. **63**, 221 (1920)
[13] A. Landé, Phys. Z. **22**, 417 (1921)
[14] A. Landé, Z. Phys. **7**, 398 (1921)
     N. Bohr, Z. Phys. **9**, (1921)
[15] H. G. J. Moseley, Phil. Mag. **26**, 1024 (1913), **27**, 703 (1914)
[16] W. Kossel, Verh. D. Phys. Ges. **16**, 898, 953 (1914), **18**, 339 (1916)
[17] A. Sommerfeld, Ann. d. Phys. **50**, 1 (1916)
[18] W. Kossel, Z. Phys. **1**, 119, **2**, 470 (1920)
     A. Smekal, Z. Phys. **4**, 26 (1920), **5**, 91, 121 (1921)
     D. Coster, Z. Phys. **5**, 139, **6**, 185 (1921)
     G. Wentzel, Z. Phys. **6**, 84 (1921)
[19] J. Franck and G. Hertz, Verh. D. Phys. Ges. **16**, 457, 512 (1914)
     N. Bohr, Phil. Mag. **30**, 394 (1915)
[20] O. Stern, Z. Phys. **7**, 249 (1921)
     W. Gerlach and O. Stern, Z. Phys. **8**, 110, **9**, 349 (1922)

8. THE STRUCTURE OF THE ATOM AND THE PROPERTIES OF THE ELEMENTS

A High Point and a Crisis

THE provisional quantum theory that was developed over the years 1913 to 1918 appeared to be capable of dealing with separable or nearly separable systems. It had led to the interpretation of series spectra in terms of the two quantum numbers n, l and to the evaluation of the spectral terms. However, the fine structure of the terms—the n, l, j model—and the behaviour of the terms in an external magnetic field still eluded explanation. We now know that the understanding of these matters was complicated by the fact that two utterly different problems were very closely intertwined: the axiomatic foundations of quantum mechanics and the 'spin' of the electron.

The interpretation of simple spectra appeared to have brought an old wish near fulfilment—the wish to understand the atom and hence the physical and chemical properties of elements: indeed, to do so in terms of very simple assumptions, viz. (a) the mass m and the charge $-e$ of the electron, (b) the charge Ze of the atomic nucleus (because it is essentially at rest its mass does not enter into atomic properties), and (c) quantum theory with its constant h. The charge number Z thus emerged as the sole parameter that determined atomic properties. In 1921–2 BOHR attempted to give an explanation of physical and chemical properties for simple spectra along these lines. In particular he tried to provide an explanation of the periodic system, which places greater emphasis on the chemistry of heteropolar bonds than of homopolar ones. BOHR's explanation was not a deduction from the assumptions given above. It used unexplained experimental facts. It was a kind of grand view. But it appeared to show the right way. The difficulties were pushed aside and the period lengths 2, 8, 18, 32 were derived from the symmetry considerations in a way which was later found to be untenable. The twin difficulties, the foundations of quantum mechanics and the spin, were joined by yet a third—that of how shells were completed, i.e., the PAULI exclusion principle.

BOHR's explanation of atomic properties was at that time felt to be a sort of 'high point in quantum theory'. Even today it may be regarded as a pinnacle of old quantum theory, a demonstration of the power of the correspondence principle. But shortly thereafter physicists became increasingly conscious of the problems bound up with this old form of quantum theory, with the result that by 1922 they began to talk of a 'crisis'. Pride and doubts were very closely intertwined. But in this chapter we shall deal with the high point.

The Periodic System and the Structure of the Atom

Even the early models of the atom were intended to provide an explanation of the chemical properties of elements. In the model suggested by Lord KELVIN and formulated by THOMSON using rings of electrons, considerations of mechanical stability led to maximum occupation numbers for the individual rings. Atoms with similar incomplete rings (these were the inner rings) suggested an explanation of the chemical homology of elements. Chemical valency might be related to easily released electrons.

Impressed by the RUTHERFORD model of the atom, BOHR was fascinated by the prospect of understanding the properties of the chemical elements in terms of a single number—the charge number of the nucleus. In the second paper of the 'trilogy' of 1913 he had assumed the existence of electron rings whose occupation numbers he fitted to experimental data. Thus Li was given an interior ring with two electrons and an external one with one electron. C was given an external ring with four electrons. The predictions did not correspond to reality, as BOHR had assumed orbits with angular momentum $h/2\pi$ per electron. Table 3 shows part of the imaginative system of occupation numbers which he put forward as a 'possible arrangement' (the rings are counted from inside). We see how the valence number guided him. Certain elements with similar properties, the 'iron group', are taken to differ in respect of their inner electrons, with the groupings of the external ones remaining unchanged. The rare earths are assumed to differ in respect of the grouping of electrons lying even further inside. We can see that the 1913 arrangement comprises important components of BOHR's later explanation. The third paper of 1913 was an attempt to interpret homopolar bonds, but it was premature. The chemical behaviour of elements which emerged in the valence number and electro-positiveness or -negativeness was derived from provisional or final maximum occupation numbers 2, 4, 8 of the rings. X-ray spectra gave further indications of a process of this kind governing the completion of rings or shells. The emission lines (Chapter 7) corresponded to transitions between energy levels that lay in the neighbourhood of $E = RchZ^2n^2$. They were thus comprehensible in terms of a nuclear charge partially screened by the electrons.

TABLE 3: BOHR'S OCCUPATION NUMBERS, 1913

1 H	1		
2 He	2		
3 Li	2	1	
4 Be	2	2	
5 B	2	3	
6 C	2	4	
7 N	4	3	
8 O	4	2	2
9 F	4	4	1
10 Ne	8	2	
11 Na	8	2	1
12 Mg	8	2	2
. .	.	.	.
18 Ar	8	8	2

In non-optical absorption there were not lines but fringes. It was possible that these corresponded to the release of an electron. Thus it was that in 1914 KOSSEL explained the absorption of X-rays not by excitation but by the separation of an electron and the emission lines by the displacement of electrons. The energy states $E(n)$ of electrons that were visible in X-ray spectra are thus occupied by electrons in the normal state of the atoms. He expressed this more clearly in 1916: the interior 'electron rings' of an atom are normally fully occupied. The valence number of other elements is equal to the number of the electrons orbiting outside such closed rings or equal to the number of the electrons needed to complete the rings. KOSSEL extended this concept to a theory of polar chemical bonds as bonds between ions that behaved like the noble gases. The formation of molecules was thus a problem of atomic structure, namely the tendency of atoms to form ions with complete shells. This concept proved to be fruitful not only for compounds such as FH, H_2O and NH_3 but also for compounds such as $[NH_4]Cl$ and for those of the 'co-ordination type' such as $[SiF_6]R_2$ or $[FeF_6]R_3$. KOSSEL was able to provide an explanation of the co-ordination number. As we have said. BOHR explained the changing valency of the elements in the neighbourhood of iron as the formation of interior groups of electrons. This point was clarified by LADENBURG in 1921.[1]

The enumeration of the ground terms of the optical spectra of the alkaline elements Li, Na, K, Rb, Cs as $n = 2, 3, 4, 5, 6$ corresponding to the idea of complete shells was known neither to BOHR in 1920 nor to SCHROEDINGER in 1921. BOHR produced it in his 1921 model of the periodic system of the elements.

Bohr Festivals

BOHR was lecturing in 1921–2 on the structure of the atom and on the physical and chemical properties of the elements. October 1921 saw the publication of a 'Lecture to the Physical Association of Copenhagen'.[2] It is inconceivable that BOHR could have delivered a lecture of such length in a single evening. The lectures that he gave at the invitation of BORN and FRANCK in June 1922 in Göttingen had far-reaching effects. Six evenings were amply taken up by his lectures; another was devoted to discussion. The content corresponded roughly to the published text, with addenda which date from April 1922, and which went a little beyond this. Among those present were SOMMERFELD, LANDÉ, PAULI and HEISENBERG. BOHR spoke, as he often did, rather indistinctly, and he was barely audible in the back rows where the younger members were obliged to sit. This merely served to increase the excitement and interest. It is impossible to recall the magic of that historic moment. All we can do is to try to recapture the most important points. Of course, this means that what does remain may well receive undue emphasis.

BOHR began by dealing with the foundations, taking the hydrogen atom and its stationary states as stages in the binding of the electrons, the X-ray spectra, and the chemical properties, before tackling the main problem. He did this by posing the question: how can an atom be formed by successive acquisition and binding of the individual electrons in the field that surrounds the nucleus? After he had explained a simple spectrum he came to his crucial review of the structure of atoms with regard to their positions in the periodic system. In some respects this turned out to be obscure and not always easy to understand. BOHR had perhaps rather misapplied his 'Platonism'.

The binding of the first electron leads to the well-known spectrum of hydrogen; $n = 1$ is ascribed to the ground term. The helium spectrum gives us some information about the binding of the second electron. Coplanar orbits are ascribed to orthohelium; skew paths to parahelium. The absence of the lowest term $n = 1$ for orthohelium was explained, in the spirit of the correspondence principle, by the fact that a plane orbit cannot be continuously transformed into one with a symmetrical arrangement of electrons. The ground state of the two-electron system must therefore be parahelium with two electrons $n = 1$, $l = 1$. Let us immediately use the notation $1s^2$ adopted later. For the third electron the Li spectrum shows that it is impossible to have $n = 1$ in the ground state. The ground state is given by $1s^2 2s$. We may exclude the possibility $1s^3$ on the basis of the correspondence principle. With a fourth electron we obtain $1s^2 2s^2$, with the sixth $1s^2 2s^4$, for a symmetrical configuration. The highly symmetrical $2s^4$ arrangement repels further electrons, so that the seventh electron in the ground state arrives in the path of next lowest energy 2p. When we reach ten electrons we obtain the very symmetrical arrangement $1s^2 2s^4 2p^4$. As the lowest s and p terms in the sodium spectrum correspond experimentally to a lower binding than in that of lithium, $n = 2$ cannot be associated with them. The sodium spectrum is a distorted hydrogen spectrum, with a few terms, $n = 1$ and $n = 2$, missing at the bottom. The eleventh electron appears in a 3s orbit. The fourteenth completes the $3s^4$ configuration which, being symmetrical, repels further electrons so that the fifteenth appears in 3p and we achieve the configuration $1s^2 2s^4 2p^4 3s^4 3p^4$ with the eighteenth electron, for Ar. The long period in the periodic system that starts here reveals an important point. The last electron in the ground state of the K and Ca atoms is bound in a 4s orbit as this has a lower energy than the 3d path. This is because it penetrates deep into the interior of the atom. If, however, the nuclear charge increases, the difference between the internal and external parts of the field gradually disappears, becoming more and more similar to a Coulomb field. At some point in this increase in nuclear charge the moment must occur when the 3d orbit is more firmly bound than the 4s orbit. We see the beginnings of this in the spectra of K and Ca^+. Although for K the 3d term is still higher than the 4p term by the time we reach Ca^+ it lies between 4s and 4p. As Sc displays a different spectrum from Al (CATALAN's research had just been published) it was assumed that from Sc onwards the 3d path was more firmly bound and that an interior 3d shell was constructed between Sc and Ni. Cu once again displays a spectrum with 4s as its fundamental term so that $4s^2$ is assumed for Zn. BOHR interpreted the period length 18 as the formation of a group of electrons $3s^6 3p^6 3d^6$ with $n = 3$. He did so because three groups each of six electrons could more easily be arranged than a grouping of the type $s^4 p^4 d^4$. The elements in the particular region of the periodic table under consideration have the properties of variable valency and of coloured and paramagnetic ions. Colour and variable valency can be explained in terms of the small energy difference between the 4s and 3d electrons. The explanation of paramagnetism is that the ions include incomplete shells of electrons. It is not until we have $2 + 8 + 18 = 28$ electrons that we reach completion with the configuration of the Ni atom. From the 29th electron onwards (Cu, Zn, Ga, Ge, As, Se, Br, Kr) once again 4s and then 4p electrons are accumulated until we reach the noble gas configuration $4s^4 4p^4$ of the Kr atom. The next long period, the fifth, is constructed in a similar way as a result of the accumulation of 5s, 4d and 5p orbits. The only difference is that the characteristic properties of 4d set in somewhat later because the 4d orbit is somewhat less firmly bound.

In the rather large period that begins with Cs (the 6s electron), the point must then be reached when the still absent four-quantum path (the 4f path) is more firmly bound than 6s and 5d with the result that an interior 4f shell is formed. This explains the occurrence of the rare earths. The period length 32 suggests the arrangement $4s^8 4p^8 4d^8 4f^8$ for the complete shell of the four-quantum paths (later developments changed this to $s^2 p^6 d^{10} f^{14}$). Four groups of six do not allow of a simple symmetrical arrangement among themselves (or so BOHR thought) but four groups of eight might perhaps do so. The seventh and last period begins with the accumulation of 7s electrons. The influence of the construction of the 5f shell actually starts rather later than that of the 4f shell in the previous period. The seventh period is thus perhaps similar to the fifth rather than to the sixth.

Table 4 gives the occupation numbers of the electron shells for the noble gases (86Nt is now called Rn).

We could summarize BOHR's formulation of the periodic system in this way: an electron shell with quantum numbers n, l is finally closed when we reach $2n$ electrons, consequently the whole group with the main quantum number n when we reach $2n \cdot n$ electrons. Provisional completion is achieved with $4 \cdot 2$ and $6 \cdot 3$ electrons. The reasons for the completion is seen in the emergence of a particularly symmetrical arrangement. The period

TABLE 4: BOHR'S OCCUPATION NUMBERS, 1922

	1 s	2 s p	3 s p d	4 s p d f	5 s p d f	6 s p d	7 s p
2 He	2						
10 Ne	2	4 4					
18 Ar	2	4 4	4 4 –				
36 Kr	2	4 4	6 6 6	4 4 – –			
54 Xe	2	4 4	6 6 6	6 6 6 –	4 4 – –		
86 Nt	2	4 4	6 6 6	8 8 8 8	6 6 6 –	4 4 –	
118 –	2	4 4	6 6 6	8 8 8 8	8 8 8 8	4 4 4	4 4

lengths are however successively not 2, 8, 18, 32 but 2, 8, 8, 18, 18 32, as the 3d, 4d, 5d... 4f, 5f... orbits are loosely bound for the outer electrons which travel in a field very different from the Coulomb field, while for the inner electrons that travel in a Coulomb-type field, the energy sequence is determined mainly by n. Thus for outer electrons 3d follows 4s, 4d follows 5s..., 4f follows 6s and 5d. The periodic system is thus explained by the application of a closure law (justified, albeit incompletely, on grounds of symmetry) and of observed energy values.

During BOHR's lecture his assistant unrolled an enormous illustration with elliptic and rosette orbits beautifully drawn on it (these were already being printed in a number of places).[3] He went on to show the representation of the periodic system given in Figure 11 in which elements with similar properties are connected by lines and the construction of interior rings is indicated by the boxes.

BOHR also suggested that the construction of the 4f shell should be completed with 70 electrons, and that thus the element 72 which had not yet been discovered and was not, as was then thought, a 'particularly rare earth' but must have properties similar to those of Zr or Th. A few months later the element was found by Dirik COSTER and Georg v. HEVESY in samples of zirconium minerals and named 72Hf, hafnium after the place where it was discovered (Hafnia being the old name for Copenhagen). BOHR's work on the periodic system was expanded in a paper by BOHR and COSTER about X-ray spectra and the periodic system, written towards the end of 1922.[4]

It examined the terms that had been deduced from the X-ray and optical spectra from the point of view of their dependence on the atomic number. 'Peculiar irregularities' in the development of this dependence showed the non-uniform behaviour of the screening of the nuclear charge and thus the formation of interior electron shells.

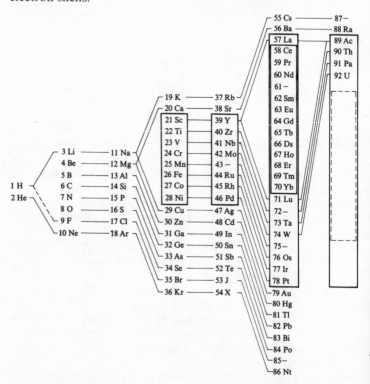

FIGURE 11: THE PERIODIC SYSTEM OF THE ELEMENTS

Light and Shade

However impressive Bohr's proposals were, it was impossible to overlook the unsolved problems. No one had yet shown convincingly how the requirement that the groups of electrons should be symmetrical could determine the provisional or final completion of such groups. (It later transpired that there were in fact other reasons for the closure.) The doublets and triplets among the terms were ignored, being explained only very generally in terms of the interaction with the other electrons and the consequent deviation from a central field. The anomalous Zeeman effects were clearly connected with the fine structure. Their analysis appeared to suggest conceptual models which did not strictly fit in with the remaining principles; so did the distribution of the multiplicities of the terms among the elements. We may regard the lecture that Born gave in the winter of 1923–4 in Göttingen as the culmination of the provisional quantum theory. He proceeded to write it down in book form.[5] It was an attempt to provide a deductive theory of the atom that would set out the limits within which the current principles held. The fourth edition of Sommerfeld's book, published at about the same time, gave equal emphasis to what could be explained and what could merely by systematized (as, for example, the anomalous Zeeman effects).

[1] W. Kossel, Verh. D. Phys. Ges. **16**, 898, 953 (1914), **18**, 339 (1916); Ann. d. Phys. **49**, 229 (1916)
  R. Ladenburg, Z. Elektrochem. **26**, 262 (1920)
[2] N. Bohr, Fys. Tidskr. **19**, 153 (1921), Z. Phys. **9**, 1 (1922), Reprinted in: Drei Aufsätze über Spektren und Atombau, Braunschweig 1922, with addenda of April 1922
[3] e.g. H. A. Kramers and H. Holst, Das Atom und die Bohrsche Theorie, Berlin 1925 Zehn Jahre Bohrsche Theorie, Sonderheft der Naturwissenschaften **11**, 533 (1923)
[4] N. Bohr and D. Coster, Z. Phys. **12**, 342 (1923)
[5] M. Born, Vorlesungen über Atommechanik I, Berlin 1925

9. TOWARDS ELECTRON SPIN

Confusing Complexity

THE foundations of what was later to be called old quantum theory were clear by about 1918. This basic quantum theory reached a crisis by 1922 as a result of its failure to predict the helium terms (Chapter 10) and through its internal inconsistencies. The crisis was not resolved until the creation in 1925 of a new version of quantum mechanics. The theory of the physical and chemical properties of the atom was also still quite fragmentary in 1922 in spite of the successes of SOMMERFELD and his pupils in the systematization of spectra and in spite of BOHR's review of the periodic system. No one had yet convincingly explained the numbers 2, 8, 18, 32 with which the shells were completed or the fine structure of the (n,l) terms. PAULI's exclusion principle and the discovery of the electron spin in 1925 provided the necessary explanation.

Today we are able to separate these two problems. But physicists were unable to do so at the time. They were not to know that the complicated properties of spectra were not necessary for an understanding of the fundamentals of the new quantum mechanics. Nor could they know that the fundamentals of quantum mechanics contributed very little to the explanation of the complicated properties of spectra. Thus it was that the two approaches obstructed each other. The aspects which were receiving the most attention were beset with a number of difficulties—the foundations of quantum theory, spin and the exclusion principle.

Let us trace the path that led to the spin. The discovery of the multiplets and the investigation of their ZEEMAN splitting were of particular importance in this context. We must therefore consider three phases of development: the multiplicity in the cases of doublets and triplets (1921), the discovery of the multiplets (1922) and its theoretical treatment and solution (1922–4); finally the explanation in terms of the PAULI principle and the spin (1925).

Anomalous Zeeman Splitting for Doublets and Triplets

By 1921 it was known that atoms with a single external electron
had doublet terms. The secondary quantum number $l = 1, 2, 3...$
(for $l = 1$ only $j = 1$) is associated with two terms which have
internal quantum number $j = l, l - 1$. Nowadays we write
$l = 0, 1, 2...$ and $j = l \pm 1/2$. Atoms with two external electrons
have simple terms and triplet terms. For the latter it was customary
to write $j = l, l - 1, l - 2$ (for $l = 1$ and $j = 1$). Nowadays we
write $j = l + 1, l, l - 1$. Atoms with three external electrons
again have doublets. Single lines display normal ZEEMAN effects.
ZEEMAN had based his work mainly on the single lines of Zn and
Cd. SOMMERFELD and DEBYE incorporated this normal ZEEMAN
effect into quantum theory in 1916,[1] by adding a magnetic
quantum number m with $\Delta m = \pm 1, 0$ and $|m| \leq l$ to the angular
momentum quantum number l and putting

$$E = mh\nu_0 \qquad \nu_0 = \frac{eB}{4\pi\mu}$$

for the energy additional to that of the electrons in the magnetic
field B (we have now written the electron mass as μ). The stipula-
tion $|m| \leq l$ has no consequences for the observed lines. Doublets
and triplets have anomalous ZEEMAN effects. It was known that
the splitting occurred in a simple ratio to the normal splitting ν_0
and depended only on the numbers l and j of the terms involved.
Basing his approach on observations by BACK, Alfred LANDÉ was
able to obtain a systematic explanation of the anomalous ZEEMAN
effects for doublets and triplets.[2] He was able to derive them from
terms that corresponded to an additional energy of

$$E = mgh\nu_0$$

where g depended on l and j. Strikingly, m was half-integral for
doublet terms

$$m = \pm 1/2, \pm 3/2... \pm j - 1/2 \quad \text{(modern notation } \pm j),$$

while for triplet terms

$$m = 0, \pm 1, \pm 2... \pm j \quad \text{(still used)}.$$

Because of the dependence of the splitting of the terms on l and j,
in distinction to the normal ZEEMAN effect, the maximum value of
n could be determined empirically. Later it was possible to replace

the unsuitable way of labelling j for doublets by a more suitable one which was lower by half. LANDÉ also gave simple formulae for g. At the end of 1921 Werner HEISENBERG attempted to provide an explanation of these phenomena.[3] If an electron joins an atom then as a result of the interaction with the electrons already there —the 'core' or 'rump' of the atom—it gives up some of its angular momentum l (in units of $h/2\pi$) namely half of one of these units to the core and retains $l - 1/2$. For two electrons outside a noble gas type shell this gave two possibilities for the core with angular momenta 0 and 1. As a result of the combination with the angular momentum of the external electron this gave rise to the singlet system in the one case and the triplet system $j = l, l - 1, l - 2$ in the other case. The explanation of the anomalous ZEEMAN effects that went with this was still not satisfactory. However, the explanation of spectra in the case of two electrons with core angular momenta (later to become the spin angular momenta) 0 and 1 and generally the assumption of an angular momentum of $\frac{1}{2}$ derived from an electron, which does not enter its orbital angular momentum, remained useful.

Multiplets

1922 was to see a major advance as a consequence of a richer supply of experimental material. While up to then only singlet, doublet and triplet terms had been known, M. A. CATALAN now discovered groups of lines in the spectrum of Mn, which could be derived from quadruple, sextuple and octuple terms. In the spectrum of Mn^+ he found quintuplet and septuplet terms. SOMMERFELD immediately established that the new multiplets followed his l, j model with $\Delta j = \pm 1, 0$. The laws of displacement and interchange both held (Chapter 7).[4] In the next few years numerous complicated spectra were resolved in terms of a derivation from multiplet terms. The theory of the multiplets was of significance for the progress of quantum theory because it substantiated theoretical principles and extended them. These principles were eventually to lead to the spin of the electron.

Using the results of CATALAN and of Hilde GIESELER and in particular BACK's research into ZEEMAN effects, LANDÉ created a theory of multiplet spectra and their ZEEMAN splitting in the spring of 1923.[5] He introduced a notation and enumeration that was followed quite generally for about two and a half years. He

described the multiplet terms by their 'apparent angular
momenta': R in units of $h/2\pi$ for the core of the atom, $L = l - 1/2$
for the external electron (corresponding to the current notation k
for this angular momentum he wrote $K = k - 1/2$) and $J = j$ for
doublets or $J = j + 1/2$ for triplets for the angular momentum of
the whole atom. The numbers R, L, J correspond in modern
notation to $S + 1/2$, $L + 1/2$, $J + 1/2$. S is now assigned to the
total spin angular momentum and L to the total orbital angular
momentum. Combining the angular momentum vectors R and L
into their resultant J according to LANDÉ gave the possibilities

$$J = L + R - 1/2, \quad L + R - 3/2 \quad \dots \quad |L - R| + 1/2.$$

This gave rise to multiplets with multiplicity $2R$ (or $2L$ if $L < R$).
In the external magnetic field the angular momentum component
m in the direction of the field took the discrete values:

$$m = J - 1/2, \quad J - 3/2 \quad \dots \quad -J + 1/2.$$

This method of forming vectors and components was once
dubbed by RUNGE 'magic multiplication'.* The terms in the
anomalous ZEEMAN effects corresponded to the additional
magnetic energies

$$E = mghv_0$$

with

$$g = 1 + \frac{J^2 - \frac{1}{4} + R^2 - L^2}{2(J^2 - \frac{1}{4})} = 1 + \frac{(J - \frac{1}{2})(J + \frac{1}{2}) + (R - \frac{1}{2})(R + \frac{1}{2}) - (L - \frac{1}{2})(L + \frac{1}{2})}{2(J - \frac{1}{2})(J + \frac{1}{2})} \tag{1}$$

Thus SOMMERFELD*'s theory with the internal quantum number*
j(Δj $= \pm 1$, 0), LANDÉ*'s explanation of this number as the total
angular momentum and* HEISENBERG*'s concept of the transfer of
the angular momentum* $\frac{1}{2}$ *per electron to the atomic core gave rise
in 1923 to the* LANDÉ R, L, J *vector model.*

LANDÉ's formula for g was given quite an exciting interpretation.[6] In the spring of 1923 PAULI was considering the terms of the

* *Translator's note:* German *Hexeneinmaleins*. In Goethe's *Faust*, I, the witch
declaims:

> *Und Neun ist Eins,*
> *Und Zehn ist keins.*
> *Das ist das Hexen-Einmal-Eins.*

anomalous ZEEMAN effect in the limiting transition to a very strong magnetic field. In the limit

$$E = (m + m_R)hv_0 = (m_L + 2m_R)hv_0$$

where m_R and m_L were the components of R and L in the direction of the field. The angular momenta R and L appeared to be individually coupled to the magnetic field and the atomic residue reacted magnetically twice over. LANDÉ himself soon explained his g formula in terms of the vector model: the vector frame composed of R and L rotates about the resultant J. This system executes a slow precession about the direction of the magnetic field (Figure 12).

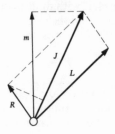

FIGURE 12: VECTOR MODEL OF THE ZEEMAN EFFECT

If R and L gave rise to magnetic moments like orbits of electrons the additional magnetic energy would be

$$E = mhv_0$$

as in the case of the normal ZEEMAN effect. A magnetic moment $\sim R$ gives an additional moment proportional to $R \cos(R,J)$ in the J direction. As a result of the precession it contributes an amount $\sim (m/J) \cdot R \cos(R,J)$. Instead of the additional amount expected in this way for g

$$\frac{R}{J} \cdot \frac{J^2 + R^2 - L^2}{2JR}$$

in the g formula the additive term

$$\frac{\tilde{J}^2 + R^2 - L^2}{2\tilde{J}^2} \qquad \tilde{J}^2 = (J - \tfrac{1}{2})(J + \tfrac{1}{2})$$

appears next to 1. The core of the atom was given a magnetic moment that was twice as large as if it were derived from orbits. And this was seen as the source of the anomaly in the ZEEMAN effect.

The 'bifurcation theorem' stated that two atomic residues could result from the ground state of an ion when an electron was acquired. This could now be expressed more generally as follows: if the ground state has the angular momentum J the acquisition of an electron leads to one core with $R = J - 1/2$ and to one with $R = J + 1/2$ (LANDÉ and HEISENBERG, 1924).[7]

We can see how by 1924 the numerous experimental results from atomic spectra confronted the theory with three peculiar facts: the bifurcation of the state of an ion on the acquisition of an additional electron, the magnetic anomaly in the angular momentum ascribed to the residue and the appearance of $\bar{J}^2$, where one had expected J^2.

We could apply the description 'peculiar ambiguity' in the case of phenomena associated with the interaction of electrons not merely to the first of these facts but also to the third. This misled HEISENBERG in 1924, inducing him to give an explanation which is a rather impressive illustration of the situation at that time[8]: one energy of interaction corresponds to two values of J. Just as earlier the quantum frequency $v = [E(n + \tau) - E(n)]/h$ had replaced the classical frequency $v = \tau v_1 = \tau \, dE/h \, dn$ let us now try to write the energy of an interaction in the form $E = dF/dJ$ and put $E = F(J + 1/2) - F(J - 1/2)$. Instead of

$$\frac{1}{J^2} = \left[\frac{d}{dJ}\left(-\frac{1}{J}\right)\right]$$

this leads to

$$\frac{1}{J-\frac{1}{2}} - \frac{1}{J+\frac{1}{2}} = \frac{1}{(J-\frac{1}{2})(J+\frac{1}{2})}$$

and instead of the formula

$$g = 1 + \frac{J^2 + R^2 - L^2}{2J^2}$$

corresponding to the dummy model (Figure 12) we obtain the actual g formula (1). We can see how HEISENBERG saw as identical two things that are today explained in completely different ways: the ambivalence of the core, afterwards explained by spin, and the expression $(J - 1/2)(J + 1/2)$ which was later explained by the eigenvalues of the square of the angular momentum. In 1924, influenced by HEISENBERG, SOMMERFELD suggested a more appropriate way of counting R, L, J instead of LANDÉ's method:

$$j_s = R - 1/2 \quad j_a = L - 1/2 \quad j = J - 1/2$$

s corresponded to the ground term (the s term), a to the excitation. Applying this, it was possible to put

$$j = j_s + j_a, \quad j_s + j_a - 1 \dots \quad |j_s - j_a|$$

and

$$m = j, j - 1 \dots - j.$$

This corresponded to the way numerical values were later assigned to S, L, J. But not very much attention was paid to it, in spite of its simplicity.

Not all physicists assigned the strange additional angular momentum to the core. At the end of 1924 PAULI gave reasons why the angular momentum of the atom could derive only from the external electrons, why in particular the single outer electron was the seat of the angular momentum for alkalis, and why the magnetic anomaly was the result of a 'peculiar and classically indescribable kind of ambiguity in the quantum properties of the emitting electron'[10] PAULI came very close to the idea of spin but he thought straight past it.

The Pauli Exclusion Principle

In his review of the properties of the elements BOHR had divided the electron numbers 2, 8, 18, 32, which completed the electron shells, in the form

$$2; \quad 4+4; \quad 6+6+6; \quad 8+8+8+8$$

TABLE 5: STONER'S OCCUPATION NUMBERS

	s	p_1	p_2	d_2	d_3	f_3	f_4
l	1	2		3		4	
j	1	1	2	2	3	3	4
Number	2	2	4	4	6	6	8
	2	6		10		14	

into sub-shells (according to *l*). In 1924 E. C. STONER suggested another partition which made it easier to understand the transition of the X-ray terms formulated in terms of *n*, *l* and *j* into the optical terms.[11] The partition no longer depended on *n* but only on *l*.

For each value of *j* according to the method of counting then in use there were at most 2*j* electrons, for each value of *l* at most $2(2l - 1)$, for each value of *n* (as given by BOHR) at most $2n^2$. STONER thus established that the number required for completion was equal to the number of components ($|m| \leqq j - 1/2$) in the magnetic field. He might have added that they were equal to the number of indivisible states.

In his crucial paper of January 1925 PAULI tried to abandon the excessively specialized conceptual model which derived the 'magnetic anomaly' from an 'ambiguity' of the core.[12] He formally ascribed it to an ambiguity in the electron, which accordingly takes two azimuthal quantum numbers (which we shall here call *l* and *j*, $\Delta l = \pm 1$, $\Delta j = 0, \pm 1$) and, in all, four quantum numbers *n*, *l*, *j* and *m*. STONER's completion numbers were then the consequence of a general rule. In an atom there is at most one electron for each set of four quantum numbers *n*, *l*, *j* and *m*. If it is present, the state is occupied. This explained the completion numbers 2, 8, 18, 32. PAULI thus also understood the absence of the lowest s term that had been expected in the triplet system of atoms with two external electrons. Both electrons had the same value of *n*, the same value of *l*, and the same value of *j* which meant that by his principle *m* would have to be different ($+\frac{1}{2}$ and $-\frac{1}{2}$); the sum 0 could correspond only to $J = 0$ and

TABLE 6: *J* VALUES OF p²

j		*m*		Σ*m*				*J*
$\frac{1}{2}$	$\frac{1}{2}$	$+\frac{1}{2}$	$-\frac{1}{2}$			0		0
$\frac{3}{2}$	$\frac{1}{2}$	$\pm\frac{3}{2}$	$\pm\frac{1}{2}$	2	1	−1	−2	2,1
		$\pm\frac{1}{2}$	$\pm\frac{1}{2}$	1	0	0	−1	
$\frac{3}{2}$	$\frac{3}{2}$	$+\frac{3}{2}$	$-\frac{3}{2}$			0		
		$\pm\frac{3}{2}$	$\pm\frac{1}{2}$	2	1	−1	−2	2,0
		$+\frac{1}{2}$	$-\frac{1}{2}$			0		

belonged to the singlet system. PAULI saw further that an atomic
state which lacked certain electron states for the completion of a
shell displayed the identical term multiplicity to a state for which
those same electron states were occupied. This was the first
appearance of the important law of symmetry between the
existence and absence of an electron, nowadays we should say of a
fermion (a particle that satisfies the FERMI statistics of indistinguishable particles that behave statistically like electrons,
Chapter 13). PAULI derived the possible J values for two equivalent p electrons with equal value of n by writing down the possi-
bilities for the values of j and (Table 6), counting $j \geq |m|$.

Explanation of the Complicated Spectra

Multiplets were found in a number of spectra in the years 1923
and 1924 and the system of terms was explained by the use of
R, L and J. It was significant that there were relatively simple
systems like Sc and Sc^+ with higher order multiplets. It was also
significant that series of terms had been discovered in the spectra
of earth alkalis whose limit suggested a higher state of the core of
the atom. Thus it was that physicists learned gradually to
distinguish between ordinary singlets, doublets and triplets which
belonged to ordinary s, p, d series and which appeared to be
constructed on the basis of the ground state of the core and
'multiplets of a higher level' which did not belong to an ordinary
state of the core. For the latter it was possible for low energy
terms to occur with high values of L (for example, 'f terms'),
which could not possibly correspond to an electron orbit of
angular momentum L.

The time was ripe for an explanation of complicated spectra,
i.e., an interpretation of them in terms of the (n,l) orbits of the
individual electrons, to be followed by their incorporation into
BOHR's periodic system. This occurred in 1925 as a result of a
number of steps which followed in swift succession.[13] H. N.
RUSSELL and F. A. SAUNDERS in Princeton, N.J. and Cambridge,
Mass. looked into the 'displaced terms' of earth alkali atoms
already mentioned and explained their L values in terms of a
vector model in which the angular momenta l_1 and l_2 of the two
electrons combined to give the resultant L. Thus, two p electrons
gave the multiplets

$$p + p \rightarrow {}^3S_1 \, {}^3P_{012} \, {}^3D_{123}.$$

This notation which was soon adopted (multiplicity on the top left, L denoted by S, P, D..., and J on the bottom right) was used by RUSSELL and SAUNDERS. In April 1925 HEISENBERG ended a review of the attempts to provide explanations, using models, 'of the ambiguity that characterizes the interaction between the core of the atom and the electron which cannot be described in mechanical terms'. According to one explanation, the electron had the effect that the state of the core appeared to be doubled; according to a second interpretation it was the state of the electron that seemed to double (PAULI's version). The two explanations really amounted to the same thing. There was yet a third solution, in which the angular momenta l of the individual electrons combined to give a resultant angular momentum, corresponding to LANDÉ's L. This was the version due to RUSSELL and SAUNDERS. It led in general to a multiplet arrangement if the interaction between this angular momentum and that ascribed to the core by the first solution, LANDÉ's R, was relatively small. Using PAULI's exclusion principle HEISENBERG was able to discover the multiplets ^{1}S ^{3}P ^{1}D ^{3}F ^{1}G for two equivalent p electrons, and ^{1}S ^{2}P ^{1}D ^{3}F ^{1}G for two equivalent d electrons.

In May of 1925 Samuel GOUDSMIT in Leiden was able to state the multiplets that corresponded to a configuration described by the l of the individual electrons. He used PAULI's principle and counted the possibilities for the various quantum numbers n, l, m_v, m_l. He partitioned the components m of j into one component m_v belonging to the core and another m_l belonging to l. Thus for p^3 he found: ^{4}S ^{2}D ^{2}P. In June 1925 Friedrich HUND in Göttingen, with a knowledge of HEISENBERG's solutions, was able to state the multiplets of the configuration with d electrons and was thus able to explain the complicated spectra of the elements

TABLE 7: MULTIPLETS OF p^2

l	m_l		m_s		$M_L = \Sigma m_l$	$M_s = \Sigma m_s$	$L\ S$
11	1	1	+	−	2	0	2 0
	1	0	±	±	1	1 0 0 −1	1 1
	1	−1	±	±	0	1 0 0 −1	
	0	0	+	−	0	0	0 0
	0	−1	±	±	−1	1 0 0 −1	
	−1	−1	+	−	−2	0	^{1}D ^{3}P ^{1}S

from Sc to Ni. He used a method of enumeration which was fundamentally the same as the GOUDSMIT method. Shortly afterwards he was able to explain the magnetic behaviour of the rare earths by deriving the ground states of their trivalent ions from a rule that asserts that of all the possible multiplets of the lowest energy configuration it was the multiplets with the highest multiplicity that lay lowest, and that the lowest of these was that which had the highest angular momentum. The enumeration used by GOUDSMIT corresponded to later group-theoretical approaches and it is thus given for two equivalent p electrons in Table 7. We have used the notation for the angular momenta which was shortly to become customary: l and s (spin) for the individual electrons; L and S for the whole atom.

In November 1925 Samuel GOUDSMIT and George UHLENBECK recognized the multiplets as a special case of more general couplings. They gave the following arrangements according to the coupling strengths: $(R_1 R_2)(L_1 L_2)$, the arrangement of the normal multiplet later to be called the RUSSELL–SAUNDERS coupling; $(R_1 L_1)(R_2 L_2)$, later to be known as J-J coupling, which gives a strong coupling of R and L for every component; $[(R_1 L_1)R_2]L_2$ and $[(R_2 R_2)L_1]L_2$ if L_2 is weakly coupled.

Electron Spin

Nowadays we tend to deduce the eigenvalue $\frac{1}{2}$ of the angular momentum (in units of $h/2\pi$) from the doublet terms of the alkalis: more generally, from the difference between the quantum numbers $l(\Delta l = \pm 1)$ and $j(\Delta j = \pm 1, 0)$ of an electron and thus from the four quantum numbers n, l, j, m given by PAULI. But at the time this would have been far from easy. Prototypes of the electron spin were given by (*a*) HEISENBERG's 1921 version, which suggested that each new electron coming into interaction with the others transferred an angular momentum of $\frac{1}{2}$ to the residue, (*b*) the 'ambivalence' of the interacting electron, and, of course, (*c*) PAULI's version. On receiving a letter from PAULI describing his formulation, at the beginning of 1925 KRONIG deduced the existence of an eigenvalue of the angular momentum of the electron. He was able to explain the doublet splitting in terms of $Z_i^2 Z_e^2$. Z_i and Z_e were effective nuclear charges for the interior and exterior respectively; for an angular momentum of the core this would be $Z_i Z_e^2$. But neither PAULI nor later BOHR would hear of this, so it was not published.[14] The reasons for this may well have been (*a*) a factor of two between the calculated and observed doublet splitting (*b*) the necessity of ascribing a doubled magnetic moment to the electron, and finally (*c*) the small magnetic moments of the atomic nuclei which could not possibly include a large component from the electrons (at that time the nucleus was assumed to be constructed of protons and electrons).

Then there was GOUDSMIT's splitting of the magnetic quantum number of the electron into two parts, one of which had the value $\pm\frac{1}{2}$, which came very close to the idea of spin. But the idea of electron spin first arose as a result of conversations between UHLENBECK and GOUDSMIT in the autumn of 1925. Looking through the published paper one recalls difficulties that were noted earlier. Doublet splitting $\sim Z_i^2 Z_e^2$ was indeed now comprehensible, but the factor 2 between prediction and experiment was still unexplained. BOHR now believed in spin; PAULI refused to do so, surely because of the factor 2 and because he just did not want to believe in an almost intuitive explanation of the 'ambivalence'. It was not until L. H. THOMAS in 1926 removed the discrepancy of the factor 2 by a more detailed consideration of the relativistic corrections[16] that even PAULI was prepared to recognize spin.

The emergence of the vector model, the PAULI exclusion principle and electron spin meant that there was a tolerably complete theory of atomic spectra. In the relatively frequent case of 'normal coupling' of the electrons the spins formed a resultant S (each contributing $s = \frac{1}{2}$) the orbital angular momenta $l_1, l_2, \ldots$ a resultant L, the angular momenta S and L together gave a resultant J. It was still not possible to understand the large energy difference between terms of different multiplicity for otherwise identical formations of electrons, for example between ^3P and ^2P for one s and one p electron. A magnetic interaction between the magnetic moments of two electrons was not sufficient, as could be seen from the orders of magnitude in question. This difficulty was removed by HEISENBERG in the context of the new quantum mechanics.

[1] A. SOMMERFELD, Phys. Z. **17**, 491 (1916)
   P. DEBYE, Phys. Z. **17**, 507 (1916)
[2] A. LANDÉ, Z. Phys. **5**, 231, 7, 398 (1921), **11**, 253 (1922)
[3] W. HEISENBERG, Z. Phys. **8**, 273 (1922)
[4] A. SOMMERFELD, Ann. d. Phys. **70**, 32 (1923)
[5] A. LANDÉ, Z. Phys. **15**, 189 (1923)

124   *The History of Quantum Theory*

[6] W. PAULI, Z. Phys. **16**, 155 (1923), **20**, 371 (1923)
   A. LANDÉ, Z. Phys. **19**, 112 (1923)
[7] A. LANDÉ and W. HEISENBERG, Z. Phys. **25**, 279 (1924)
[8] W. HEISENBERG, Z. Phys. **26**, 291 (1924)
[9] A. SOMMERFELD, Ann. d. Phys. **73**, 209 (1924)
[10] W. PAULI, Z. Phys. **31**, 373 (1925)
[11] E. C. STONER, Phil. Mag. **48**, 719 (1924)
[12] W. PAULI, Z. Phys. **31**, 765 (1925)
[13] H. N. RUSSELL and F. A. SAUNDERS, Astrophys. Jn. **61**, 38 (1925)
   W. HEISENBERG, Z. Phys. **32**, 841 (1925)
   S. GOUDSMIT, Z. Phys. **32**, 794 (1925)
   F. HUND, Z. Phys. **33**, 345, 855 (1925)
   S. GOUDSMIT and G. E. UHLENBECK, Z. Phys. **35**, 618 (1926)
[14] cf. W. PAULI Memorial Papers (edited by M. FIERZ and V. F. WEISSKOPF), New York 1960; see the contributions of R. KRONIG and B. L. VAN DER WAERDEN
[15] G. E. UHLENBECK and S. GOUDSMIT, Naturwiss, **13**, 953 (1925), Nature **117**, 264 (1926) cf. S. GOUDSMIT in Physikertagung 1965, Frankfurt (Main), p. 1
[16] L. H. THOMAS, Nature **117**, 514 (1926)

10. THE MATRIX VERSION OF QUANTUM MECHANICS[1]

The Failure of the Old Quantum Theory

BY 1923 it was quite clear to physicists engaged in quantum mechanics that the version in use was not fully capable of describing the properties of the atom. It was felt to be an internal inconsistency that it was necessary to introduce a classical orbit with frequency $v(E)$ in order to calculate a stationary state, while in every other respect this state contained the quantum frequencies $v = [E(n + \tau) - E(n)]/h$. As EPSTEIN had already seen in 1921 (Chapter 6), not only absorption and emission but also the dispersion of light had to be governed by the quantum frequencies. Moreover, it was possible to apply the old quantum theory only to periodic or multiply-periodic systems. Was the quantum of action h to have no part to play in non-periodic phenomena? Then there was the failure of the old quantum theory to predict the states of the helium atom. It was also regarded as a failure of quantum theory that it could not provide a satisfactory explanation of the anomalous ZEEMAN effect or of the multiplets. It appeared to be incapable of dealing properly with the interaction between electrons. Some kind of 'non-mechanical cause' was felt to be an essential feature, while in fact the spin of the electron was responsible.

At the end of 1922 KRAMERS attempted to predict the ground state of the helium atom. He derived an incorrect value of the work required for ionization. The calculation for excited helium states was attempted by BORN and HEISENBERG in the spring of 1923. Although the system was ideally suited to a perturbation calculation, their results did not agree with the spectroscopic terms. They saw two ways out of this dilemma. Either the quantum conditions were false or the mechanical equations were no longer valid even for the stationary state. At the same time J. H. van VLECK made predictions of the energy levels of the helium atom which conflicted with experimental evidence.[2] The dispersion of light atoms became the centre of interest in 1924. It smoothed the path to quantum mechanics.

The major centres of research were Copenhagen and Göttingen. BOHR and KRAMERS lived in Copenhagen. PAULI spent some time there in 1922, and HEISENBERG paid a few short, and some longer, visits from 1924 onwards. SLATER was there for a few months in 1924. At Göttingen BORN succeeded to the chair of theoretical physics in 1921, at the same time as FRANCK was appointed to a chair in experimental physics. From 1921 to 1922 PAULI was BORN's assistant. HEISENBERG spent the winter of 1922–3 in Göttingen, wrote his thesis in Munich on a problem in hydrodynamics in the summer of 1923, and was in Göttingen from the autumn of 1923 onwards, apart from a few breaks. The influences on HEISENBERG are worth noting: SOMMERFELD, whose approach was mathematical and pragmatic, BORN, whose interests were in the mathematical and systematic aspects, and BOHR, the synthesizer, the most profound of them all, with his deeper, more philosophical approach.

At that time there were doubts as to the validity of the quantum conditions $\oint p_k \, dq_k = hn_k$. They were taken to be one way of satisfying the correspondence principle, which was felt to be rather superior. A different and more correct way of 'sharpening' the correspondence principle was sought. One possible loosening of the quantum conditions was given by recalling that $\Delta\Phi = h$, for one degree of freedom, which also allowed the possibility $\Phi = h(n + \alpha)$ for a constant α. PLANCK had assumed in 1911–12 that

$$E = h\nu(n + \tfrac{1}{2})$$

for the harmonic oscillator, i.e., E was the average value of the energies in the phase regions $\Delta\Phi = h$. This led to a 'zero point energy' $h\nu/2$, which, however, initially had no verifiable consequences. It was first revealed by band spectra of molecules in which one atom could be replaced by an isotope. The frequencies can be roughly broken down into those due to the motion of electrons ν_e, the oscillation ν_o and the rotation ν_r of molecules:

$$\nu = \nu_e + \nu_o + \nu_r.$$

Although in the transition to an isotope the frequency ν_e of the electron motion remained unchanged, the constants ν_1 and ν_2 in the oscillation frequency

$$\nu_o = \nu_1(n_1 + \alpha) - \nu_2(n_2 + \alpha)$$

of the oscillators in the upper and lower electron state depended on the atomic masses, which were thus different for isotopes. This meant α could be determined. Thus R. S. MULLIKEN discovered the value $\alpha = \frac{1}{2}$ for the oscillations of the BO molecule.

For the rotator $\Phi = h(n + \alpha)$ leads to an expression $E = B(n + \alpha)^2$ for the energy and to frequencies given by

$$v(n+1,n) = 2B(n+\alpha+\tfrac{1}{2})$$

$\alpha = \frac{1}{2}$ was decided on, after a few other attempts at explaining them, from the patterns of the rotation lines of a typical vibration-rotation spectrum of a biatomic molecule (Figure 13). The series found for He_2 was explained by putting $\alpha = \frac{1}{4}$ (Figure 14).

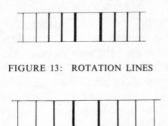

FIGURE 13:   ROTATION LINES

FIGURE 14:   ROTATION LINES FOR He_2

The 'Virtual Field' of an Atom

The connection between the electromagnetic radiation in space and the stationary states of an atom was still unexplained. EINSTEIN's radical solution had been to introduce quanta of light with energy hv and momentum hv/c, which caused the atom to go into a different energy state when they were emitted or absorbed. It was, however, incapable of disposing of the phenomenon of the interference of light. There was accordingly as yet still little sympathy with the idea of light quanta. In the course of a comprehensive discussion of the difficulties in November 1922 BOHR expressed himself as being dissatisfied with the assumption of light quanta in view of the interference phenomena.[3] He ascribed 'latent radiation reactions' to the stationary states of the atoms. These were intended to explain the probability of energy exchange in terms of the radiation. As he was obliged to assume coherent phase relations as between the incoming and outgoing waves for dispersion phenomena, he was drawn to the view that atoms reacted to radiation like a system of harmonic oscillators with the quantum frequencies $v(n + \tau, n)$ and $v(n, n - \tau)$. He did not give a detailed derivation of energy and momentum. C. G. DARWIN (1922–3) assumed classical physics to be valid for aether phenomena and thus thought in terms of the emission of a continuous light wave by an excited atom. However, within the atom quantum theory must hold, with transitions between discrete energy values.[4] It was then thought that this meant that the law of the conservation of energy could hold only in the mean. The model developed by J. C. SLATER at the beginning of 1924 was related to BOHR's 1922 view. He assumed that an atom in a stationary state was surrounded by a 'virtual radiation field', and in fact by the possible frequencies of emission or absorption of light. This virtual field would determine the probability of quantum transitions of the atom itself and of other atoms. Here too the discussions between SLATER, BOHR and KRAMERS led to the view that the laws of conservation of energy and momentum could only hold in the mean.[5]

The BOHR-KRAMERS-SLATER view that the laws could have only statistical validity was soon—we might say fortunately—disproved. In the summer of 1924 Walter BOTHE and Hans GEIGER had announced their intention of carrying out an experiment which would determine whether the COMPTON effect resulted in the simultaneous emission of the deflected light quantum when it impinged on an electron. April 1925 saw the proof that electron counters and light counters reacted simultaneously. In the meantime COMPTON and SIMON had also found, towards the end of 1924, using stereoscopic pictures, that in the cloud chamber the trace of the electron and the action of the light quantum tallied with the laws of energy and momentum.[6]

Although the view that the laws had only statistical validity soon faded, the concept of the 'virtual oscillators' with frequencies $v(n + \tau, n)$ and $v(n, n - \tau)$ remained in existence and showed itself to be useful in connection with the theory of the dispersion of light. The frequencies $v(n, l)$ and the corresponding intensities were present in the stationary state in a way that could not be imagined intuitively.

Dispersion

While the emission and absorption of light seemed to be connected with a quantum jump, its dispersion did not. However, in classical theory, absorption and dispersion were closely linked. Both depended on the facts that it was possible to displace electric charges inside atoms and that an atom in an external electric field E required an electric dipole

$$p = \alpha E.$$

Neglecting the attenuation of the radiation we obtain for α:

$$\alpha = \frac{e^2}{4\pi^2 m} \sum \frac{f_i}{v^2 - v_i^2}$$

for electrons that are quasi-elastically bound, with frequency v_i, and oscillating in sympathy with an electromagnetic wave of frequency v. In quantum theory the v_i had to be the absorption frequencies and the f_i had to depend on the (EINSTEIN) probabilities of transition between stationary states, in order to preserve the relationship between absorption and dispersion (LADENBURG, 1921). So that the classical formulae should hold asymptotically for highly excited atoms, the frequencies of emission must also appear in the dispersion formula. This meant that

$$\alpha = \frac{e^2}{4\pi^2 m} \left[\sum \frac{f_i^{(a)}}{v^2 - v_i^{(a)2}} - \sum \frac{f_i^{(e)}}{v^2 - v_i^{(e)2}} \right] \tag{1}$$

(KRAMERS, March 1924). Here [a] and [e] denote absorption and emission. Van VLECK and BORN expanded these correspondence-type arguments.[7]

Van VLECK started out from the correspondence between the individual Fourier components of the classical motion and the transition to another quantum state and showed (June 1924) that the difference between the intensities of absorption ($n_k \rightarrow n_k + \tau_k$) and the induced emission corresponded to the intensity of the classical absorption of a frequency ($n_k \rightarrow n_k - \tau_k$). In this way he was able to explain KRAMERS's dispersion formula (1).

BORN realized in June 1924—as BOHR had already done in November 1922—that similar problems arose in perturbation theory. If atoms interacted by means of radiation with the quantum frequencies $v(n + \tau, n)$ and not with the classical frequencies that had been used to calculate the stationary states, $\tau v(E)$, this must also hold for the interaction between the various parts of an atom. It was thus impossible to express this interaction in terms of a classical perturbation calculation. The perturbation method had in fact to be modified quite considerably. It contained the expressions

$$\sum_k \tau_k \frac{\partial}{\partial I_k} \left(\frac{C^2_{\tau_1 \tau_2 \dots}}{\tau_1 v_1 + \tau_2 v_2 + \dots} \right).$$

The BOHR method of converting a classical quantity into a quantum one,

$$\sum_k \tau_k \frac{\partial E}{\partial I_k} \to \frac{1}{h} \left[E(n_1 + \tau_1, n_2 + \tau_2 \dots) - E(n_1, n_2 \dots) \right]$$

could be extended to arbitrary functions $F(I)$, omitting the summation:

$$\tau \frac{dF}{dI} \to \frac{1}{h} \left[F(n + \tau) - F(n) \right].$$

For quantities F that also depended on τ, whose quantum counterpart was thus associated with a transition $(n + \tau, n)$, it seemed reasonable to perform the transition according to

$$\tau \frac{d}{dI} F(I, \tau) \to \frac{1}{h} \left[\Gamma(n + \tau, n) - \Gamma(n, n - \tau) \right]$$

where $\Gamma(n + \tau, n)$ corresponded to the classical quantity $F(I, \tau)$. In perturbation theory and in dispersion theory the quantities that appeared should accordingly be converted in the following way:

$$\tau \frac{d}{dI} \left(\frac{C^2_\tau}{\tau v} \right) \to \frac{\Gamma(n + \tau, n)}{h v(n + \tau, n)} - \frac{\Gamma(n, n - \tau)}{h v(n, n - \tau)}.$$

The summations which BORN in fact gave are once more omitted. In July 1924 KRAMERS formulated his dispersion theory in exactly the same way and suggested that only observable quantities should now be admitted into the theory. The detailed formulae were given by KRAMERS and HEISENBERG in December 1924.[8] HEISENBERG had also consulted BORN in formulating his conversion formula.

The quantum f_i must correspond asymptotically to the numbers of electrons involved in absorption and dispersion. Thus laws were obtained for the sums of the f_i. They were drawn up by KUHN in Copenhagen and THOMAS in Breslau in May and June 1925.[9] They contained transfers of classical expressions into quantum theory which were eventually included in the rigorous quantum mechanics soon to be developed by HEISENBERG.

In 1923 SMEKAL had taken the idea of light quanta as the basis of his view that there could occur not only the coherent scattering of a frequency v but also an incoherent scattering with frequencies $v \pm v_s$ where v_s is an emission or absorption frequency of the scattering system.[10] These frequencies are now contained in the KRAMERS-HEISENBERG dispersion formula. They were discovered experimentally in 1928 by Chandrasekhara Venkata RAMAN. This 'RAMAN effect' proved to be very useful for the measurement of the vibration frequencies of molecules and thus also for the explanation of molecular forces.

Intensity Rules

In classical theory the intensities of radiation of an atom were determined by its state, and they could be calculated by the Fourier series of a co-ordinate (or sum of co-ordinates)

$$x = \sum_\tau x_\tau(E)\, e^{i\tau\omega(E)t} \qquad (\omega = 2\pi v).$$

In quantum theory they were associated with a transition and a frequency $\omega(n, n - \tau)$. The correspondence principle could thus provide only approximate values of the intensities. Improvements of the correspondence principle as a result of precise intensity ratios were made by H. B. DORGELO, H. C. BURGER, L. S. ORNSTEIN, Helmut HÖNL, Samuel GOUDSMIT and Ralph KRONIG.[11] Measurements by DORGELO had given the intensity ratio 2:1 for the s-p doublets of alkalis and the formula 5:3:1 for the s-p triplet of earth alkalis. On the strength of a guess by SOMMERFELD, DORGELO was also able to prove the intensity ratios for s-p transitions in systems of higher multiplicities to be equal to the ratios of the J values in LANDÉ's enumeration, which are now written as $J + \frac{1}{2}$. ORNSTEIN and BURGER saw in this the relationship for the statistical weights ($2J + 1$ in modern notation). BURGER and DORGELO went on to give intensity rules for line groups where both terms were simple. The laws stated that the sums of the intensities of the line groups, whether derived from one term or combining into one term, behaved according to their J values. This gave the unique intensities for a doublet-doublet combination $l \rightarrow l + 1$ shown in Table 8.

TABLE 8:   INTENSITIES IN A JUXTAPOSED DOUBLET

	$j=l$	$j=l+1$	Sum
$j=l$	1	$(l+1)(2l-1)$	$l(2l+1)$
$j=l-1$	$(l-1)(2l+1)$	0	$(l-1)(2l+1)$
Sum	$l(2l-1)$	$(l+1)(2l-1)$	

The rule on its own did not give unique formulae for a triplet-triplet transition $l - \frac{3}{2}, l - \frac{1}{2}, l + \frac{1}{2} \rightarrow l - \frac{1}{2}, l + \frac{1}{2}, l + \frac{3}{2}$. Simple expressions could however be guessed. ORNSTEIN and BURGER gave intensity ratios for ZEEMAN components in simple cases by prescribing the ratios of the sums of the intensities for the individual directions of polarization. HÖNL was able to produce intensity laws for all the ZEEMAN components from the obvious sharpening of the intensities that came from classical considerations. GOUDSMIT and KRONIG gave the same rules by adding to the previous requirements that of a quadratic dependence of intensity on the magnetic quantum number.

The dispersion formula was crucial in preparing the correct improvement of the correspondence principle. The summation laws given by KUHN and THOMAS also helped in this respect. It was later easy to derive the intensity formulae from the new quantum mechanics and they thus helped to confirm the validity of the new theory.

Heisenberg's Reinterpretation of Quantum Theory

The first strictly valid version of quantum mechanics that was both logical and capable of generalization was provided by HEISENBERG's paper of July 1925.[12] He accused the previous formal rules of quantum theory of containing magnitudes such as position and frequency, which were not in fact observable. The laws were illogical in so far as atoms reacted to periodically changing fields not with the properties of classical orbits but with the quantum frequencies. HEISENBERG therefore attempted to create a new quantum mechanics using only observable quantities. He saw the beginnings of this in BOHR's frequency condition and in KRAMERS' dispersion theory.

Instead of the classical frequencies $\tau\omega$, the frequencies $\omega(n + \tau, n)$ must appear in the quantum expression for a co-ordinate as a function of time $x(t)$. The latter frequencies must satisfy the combination principle for spectra. This implies the transformation law:

$$\rho\omega(n) + (\tau - \rho)\omega(n) = \tau\omega(n) \rightarrow \omega(n + \rho, n) + \omega(n + \tau, n + \rho) = \omega(n + \tau, n). \quad (2)$$

(The classical formula will always appear on the left from now on, with the corresponding quantum expression on the right.) An amplitude is likewise observable: it is classically expressed by a term in a Fourier series. This suggests the transformation

$$x_\tau(n) e^{i\tau\omega(n)t} \rightarrow x(n + \tau, n) e^{i\omega(n + \tau, n)t}. \quad (3)$$

The way this is assembled classically into a motion

$$x(n, t) = \sum_\tau x_\tau(n) e^{i\tau\omega(n)t}$$

has no counterpart in quantum theory. The set of all the classical Fourier terms does, however, correspond to the set of all the quantities

$$x(n, l) e^{i\omega(n, l)t}.$$

As the classical quantity x is real, we thus have $x_{-\tau}(n) = x_\tau(n)^*$ and we must assume that

$$x(l, n) = x(n, l)^*.$$

We must use entities of this kind in our calculations from now on. As multiplication of classical quantities $xy = z$ is carried out according to the scheme

$$\sum_\rho x_\rho e^{i\rho\omega t} \cdot \sum_{\tau - \rho} y_{\tau - \rho} e^{i(\tau - \rho)\omega t} = \sum_\tau z_\tau e^{i\tau\omega t} \qquad \sum_\rho x_\rho y_{\tau - \rho} = z_\tau$$

consideration of (2) suggests the transformation

$$\sum_\rho x_\rho y_{\tau-\rho} = z_\tau \to \sum_k x(n,k)\, y(k,l) = z(n,l).\tag{4}$$

It need not be true that $xy = yx$ for the product of these sums. Under certain circumstances a classical product xy must be replaced by the symmetrical expression $(xy + yx)/2$.

The previous quantum condition,

$$\oint p\, dx = \oint m\dot{x}^2\, dt = hn$$

written in terms of Fourier series, now amounts to

$$2\pi \dot{m} \sum_\tau \tau^2 |x_\tau(n)|^2\, \omega(n) = hn$$

where the time-dependent terms in $\dot{x}^2$ disappear if we integrate over a period. This expression is not really suitable for transformation. But the somewhat more general condition

$$\frac{d}{dn} \oint p\, dx = h$$

with the Fourier series

$$2\pi m \sum_\tau \tau \frac{d}{dn}(\tau\omega |x_\tau|^2) = h\tag{5}$$

can be transformed according to the rules of KRAMERS' dispersion theory:

$$(5) \to 2\pi m \sum_{\tau=-\infty}^{\infty} \left[\omega(n+\tau,n)|x(n+\tau,n)|^2 - \omega(n,n-\tau)|x(n,n-\tau)|^2\right] = h$$

or

$$(5) \to 4\pi m \sum_{\tau>0} \left[\omega(n+\tau,n)|x(n+\tau,n)|^2 - \omega(n,n-\tau)|x(n,n-\tau)|^2\right] = h.\tag{6}$$

This formula had already appeared in the summation law given by KUHN and THOMAS.

The application to the harmonic oscillator given by HEISENBERG is simple, because the only cases that can arise are $\tau = \pm 1$, and so all the ω are equal. Formula (6) now reads:

$$4\pi m\omega\left[|x(n+1,n)|^2 - |x(n,n-1)|^2\right] = h.$$

Using the abbreviation $\hbar = h/2\pi$, which did not, however, become customary until some years later, $2m\omega|x(n+1,n)|^2$ increases by $\hbar$ if n is increased by one. As we must have $|x(0,-1)|^2 = 0$ it follows that:

$$2m\omega|x(n,n-1)|^2 = hn. \tag{7}$$

The energy calculated according to quantum theory is represented by the set of all the $E(n,l)$,

$$\left(\frac{m}{2}\dot{x}^2 + \frac{m\omega^2}{2}x^2\right)_{nl} = m\omega^2[|x(n+1,n)|^2 + |x(n,n-1)|^2]\delta_{nl}. \tag{8}$$

The terms $n \neq l$ which arise in the kinetic and potential energies disappear when we add. The energy is thus a quantity with which is associated a state (n) and not a transition (n,l), and it is not time-dependent. Working this out from (7) and (8) we obtain

$$E(n) = \frac{h}{2\pi}\omega(n+\tfrac{1}{2}) = \hbar\omega(n+\tfrac{1}{2}).$$

Using $\oint p\,dx = hn$, $\hbar\omega n$ had earlier been obtained. HEISENBERG still calculated the energy values of a simple type of anharmonic oscillator by an approximation.

HEISENBERG's calculation for the harmonic oscillator was later to become an important part of quantum field theory, where linear fields are resolved into harmonic oscillators (Chapter 15).

The Development of Quantum Mechanics

In September 1925 Max BORN and Pascual JORDAN attempted to develop a mathematical theory based on HEISENBERG's physical ideas, initially for one degree of freedom.[13] They recognized HEISENBERG multiplication as matrix multiplication. Thus physical quantities were represented by matrices whose rows and columns (initially) corresponded to the energy values. The quantum condition was expressed rather more generally without using $p = m\dot{q}$. The classical quantity

$$\frac{1}{2\pi}\frac{d}{dn}\oint p\dot{q}\,dt = i\sum_\tau \tau \frac{d}{dn} q_\tau p_{-\tau}$$

expressed in Fourier series, and translated into the language of quantum theory, gave the equation

$$i\sum_{k}\left[p(n,k)\,q(k,n)-q(n,k)\,p(k,n)\right]=\hbar$$

where the left hand side is the general diagonal element of the matrix

$$i(pq-qp).$$

The impact of the remainder of BORN and JORDAN's paper was reduced as the result of a rather inappropriate definition of the derivative of a function. The paper that appeared in November 1925[14] as the result of the collaboration of BORN, HEISENBERG and JORDAN (known in German as the '*Dreimännerarbeit*') enabled the derivatives with respect to the canonical variables p and q to be converted into algebraic expressions, by using a more appropriate definition of differentiation which preserved the product rule. By introducing the equation

$$i(pq-qp)=\hbar \qquad (9)$$

as the fundamental relationship they were easily able to derive the formulae

$$\frac{\hbar}{i}\,\frac{\partial f}{\partial p}=fq-qf \qquad \frac{\hbar}{i}\,\frac{\partial f}{\partial q}=pf-fp$$

for all rational functions. These enabled them to transform the canonical equations of motion into the form

$$\frac{\hbar}{i}\,\dot{q}=Hq-qH \qquad \frac{\hbar}{i}\,\dot{p}=Hp-pH$$

and (for all rational functions $g(p,q)$) into the general form

$$\frac{\hbar}{i}\,\dot{g}=Hg-gH. \qquad (10)$$

The authors now introduced the concept of canonical transformations into the new mechanics. A transformation with a matrix S was regarded as canonical if it left the commutation relation (9) invariant. A transformation of this kind was

$$\bar{p}=SpS^{-1} \qquad \bar{q}=SqS^{-1} \qquad A(\bar{p},\bar{q})=SA(p,q)S^{-1}.$$

Real quantities A were represented in terms of matrices with the property $A_{nl} = A_{ln}^*$; by what are called 'Hermitian' matrices. If we agree on the convention $(A^*)_{nl}$ for A_{ln}^*, the equation $A = A^*$ characterizes Hermitian matrices. So that a matrix A may remain Hermitian under canonical transformations, we must have $S^{-1} = S^*$. To solve a problem in quantum mechanics now meant starting with some suitable canonical matrices p and q (whose rows and columns did not in general correspond to the energy states of the system) and then transforming the quantity H into a diagonal matrix by a canonical transformation. The perturbation method, which put

$$S = S_0 + \lambda S_1 + \lambda^2 S_2 + \cdots$$

also fitted into this programme. BORN, HEISENBERG and JORDAN also saw that the search for a diagonal matrix corresponded to the principal axis transformation of an Hermitian form

$$\sum A_{nl} x_n^* x_l \qquad A_{nl} = A_{ln}^*.$$

Thus it was possible to use the theory of quadratic forms of infinitely many variables that had been developed by HILBERT and HELLINGER, even if the assumptions made by HILBERT and HELLINGER were rather too narrow for quantum mechanics.

In the winter of 1925–6 BORN was lecturing in Cambridge, Mass.[15] His lectures gave a lively and stimulating description of the Göttingen theory, its achievements and the hope that it gave. DE BROGLIE's wave concept was not yet accepted, and the SCHROEDINGER equation was not yet known.

As soon as he saw HEISENBERG's paper of July 1925, P. A. M. DIRAC likewise set about developing a mathematical theory.[16] It was completed at almost the same time as the work of BORN, HEISENBERG and JORDAN and in some respects it is more general. DIRAC deduced from the properties of differentiation (sum and product rules) that they could be expressed in the form

$$\frac{\mathrm{d}x}{\mathrm{d}v} = xa - ax.$$

In the case of $v = t$(time), a was the diagonal matrix of the spectroscopic terms. He saw the correspondence between classical and quantum mechanics in the fact that $xy - yx$ behaved asymptotically like the classical expression

$$i\hbar \left(\frac{\partial x}{\partial q} \frac{\partial y}{\partial p} - \frac{\partial y}{\partial q} \frac{\partial x}{\partial p} \right). \tag{11}$$

He took as his fundamental postulate

$$xy - yx = i\hbar[x,y]$$

where $[x,y]$ denoted the value of the classical expression in the bracket in (11). A particular instance of this was

$$pq - qp = \hbar/i.$$

It was not difficult to see how this could be generalized to several degrees of freedom. The postulates

$$\left.\begin{array}{l} p_k p_l - p_l p_k = 0 \\ q_k q_l - q_l q_k = 0 \\ p_k q_l - q_l p_k = \delta_{kl}\hbar/i \end{array}\right\} \qquad (12)$$

are given both in the BORN-HEISENBERG-JORDAN paper and in that of DIRAC. The former paper also proved a number of theorems about the angular momentum. From the commutation relations (12) the authors deduced those for the components P_x, P_y and P_z of the angular momentum vector:

$$P_x P_y - P_y P_x = \frac{\hbar}{i} P_z$$

and also the eigenvalues $J, J - 1,..., -J$ (multiplied by $\hbar$) of one component and the eigenvalues $J(J + 1)$ (times $\hbar^2$) of the square of the angular momentum. By this method they were able to verify intensity rules that had already been discovered. In January 1926 DIRAC was to develop a simple technique for dealing with angular and action variables.[17] Compared with working with classical quantities it was quite difficult to handle matrices. Thus it required great skill to calculate the energy values of an atom with a single electron, as PAULI did in January 1927.[18] He established the commutation relations for the co-ordinates x, y, z, for the distance r, for the components of the momentum p and the angular momentum P. He used a vector

$$A = \frac{1}{Zme^2} P \times p + \frac{x}{r}$$

with the direction of the major axis of the elliptic path and the magnitude of the eccentricity. LENZ had used this vector in making predictions for the H-atom. He 'diagonalized' E, P_z and P^2 and finally obtained the energy eigenvalues $E(n)$. When calculating the energy involved in the STARK effect using parabolic co-ordinates he diagonalized E, P_z and A_z. Perpendicular electric and magnetic fields now no longer generated the difficulties that had earlier resulted from the fact that certain stationary states (for example paths which passed through the nucleus) had had to be excluded. Simultaneously with PAULI, DIRAC also published predictions for the H atom, albeit not so conclusive in nature.[17]

Limbering up

DIRAC's realization that the essence of quantum theory lay in the algebraic relationships between the 'q-numbers', and not in the particular representation in terms of matrices, may be regarded as a generalization of the matrix form of quantum mechanics. BORN, HEISENBERG and JORDAN had indicated the connection with HILBERT's work, and this pointed to a connection with systems of orthogonal functions. Simultaneously LANCZOS also derived a connection with systems of functions (December 1925) by assigning a matrix of the form

$$a_{kl} = \int f(x,y)\varphi_k(x)\varphi_l(y)\,dx\,dy$$

to what he took to be a real system of functions φ_k. However, he made no use of the freedom of choice of this system of functions. Two related papers certainly contained elements of what was later to be called transformation theory, but they did not lead to any success.[19]

In January 1926 BORN and WIENER attempted to replace the matrices by linear operators but failed to find the method of replacing the momentum p by the operator $\hbar d/i\,dq$.[20] After attending lectures given by BORN in America and reading the first of SCHROEDINGER's papers, ECKART saw that it was possible to satisfy the commutation relation (9) with this operator in place of p. He thus brought the SCHROEDINGER equation into quantum mechanics. However, he was obliged to acknowledge in a footnote added in proof that SCHROEDINGER had already demonstrated the equivalence of the matrix form and the SCHROEDINGER equation in a second paper.[21]

---

[1] cf. The introduction to Sources of Quantum Mechanics, by B. L. VAN DER WAERDEN, Amsterdam 1967

[2] H. A. KRAMERS, Z. Phys. **13**, 312 (1923)
M. BORN and W. HEISENBERG, Z. Phys. **16**, 229 (1923)
J. H. VAN VLECK, Phys. Rev. **21**, 372 (1923)
[3] N. BOHR, Z. Phys. **13**, 117 (1923)
[4] C. G. DARWIN, Nature **110**, 841 (1922, **111**, 771 (1923)), Proc. Nat. Ak. Sci. **9**, 25 (1923)
[5] J. C. SLATER, Nature **113**, 307 (1924), **116**, 278 (1925)
N. BOHR, H. A KRAMERS and J. C. SLATER, Phil. Mag. **47**, 785 (1924), Z. Phys. **24**, 69 (1924)
[6] W. BOTHE and H. GEIGER, Z. Phys. **26**, 44 (1924), **32**, 639 (1925), Naturwiss. **13**, 440 (1925)
A. H. COMPTON and A. SIMON, Phys. Rev. **25**, 306 (1925)
[7] R. LADENBURG, Z. Phys. **4**, 451 (1921)
H. A. KRAMERS, Nature **113**, 673 (1924)
J. H. VAN VLECK, Phys. Rev. **24**, 330 (1924)
M. BORN, Z. Phys. **26**, 379 (1924)
[8] H. A. KRAMERS, Nature **114**, 310 (1924)
H. A. KRAMERS and W. HEISENBERG, Z. Phys. **31**, 681 (1925)
[9] W. KUHN, Z. Phys. **33**, 408 (1925)
W. THOMAS, Naturwiss. **13**, 627 (1925)
[10] A. SMEKAL, Naturwiss. **11**, 873 (1923)
[11] H. B. DORGELO, Z. Phys. **22**, 170 (1924)
H. C. BURGER and H. B. DORGELO, Z. Phys. **23**, 258 (1924)
L. S. ORNSTEIN and H. C. BURGER, Z. Phys. **24**, 41, **28**, 135, **29**, 241 (1924), **31**, 355 (1925)
H. HÖNL, Z. Phys. **31**, 340 (1925)
S. GOUDSMIT and R. DE L. KRONIG, Naturwiss. **13**, 90 (1925)
R. DE L. KRONIG, Z. Phys. **31**, 885 (1925)
[12] W. HEISENBERG, Z. Phys. **33**, 879 (1925)
[13] M. BORN and P. JORDAN, Z. Phys. **34**, 858 (1925)
[14] M. BORN, W. HEISENBERG, P. JORDAN, Z. Phys. **35**, 557 (1926).
[15] M. BORN, Probleme der Atomdynamik, Berlin 1926
[16] P. A. M. DIRAC, Proc. Roy. Soc. **109**, 642 (1926)
[17] P. A. M. DIRAC, Proc. Roy. Soc. **110**, 561 (1926)
[18] W. PAULI, Z. Phys. **36**, 336 (1926)
[19] C. LANCZOS, Z. Phys. **35**, 812, **36**, 401, **37**, 405 (1926)
[20] M. BORN and N. WIENER, Z. Phys. **36**, 174 (1926)
[21] C. ECKART, Proc. Nat. Ak. Sci. **12**, 473 (1926), Phys. Rev. **28**, 711 (1926)

Translator's note: Many of these papers are available in English in Sources of Quantum Mechanics edited by B. L. van der WAERDEN.

11. THE MATTER WAVE AND
THE SCHROEDINGER EQUATION

Duality

ONE way of explaining quantum theory in physical terms these days consists in regarding it as a completely non-intuitive unification or two intuitive pictures, i.e., classical particles and classical waves of fields. This indeed holds for both matter and light. The classical theory of particles is actually distorted to the point where the particles might just as well be waves. This is achieved by the quantization of periodic motion which leads to discrete values of the energy and occasionally to discrete values of angular momentum. It can alternatively be done by working in terms of mathematical models which satisfy the commutation relation $i(pq - qp) = \hbar$. The intuitive theory of matter waves and of fields of matter is modified to the point where there is room for the idea of particles. This arises from a kind of quantization of the matter distribution which yields particle numbers as quantum numbers. In the case of light waves the modification of the intuitive field picture is achieved by the quantization of characteristic modes of vibration of the electromagnetic field. The modification of the particle concept emerges in a particular form of statistics satisfied by light quanta. The basis of this version of quantum theory is thus the 'duality' of light and matter. A quantum 'mechanics' is possible only for matter. Light is first and foremost part of relativity, where there is no counterpart to the theory of the mechanics of systems of many particles. It is possible to carry out the above procedure for the quantum mechanics of matter. The quantization of particle mechanics and that of intuitive field theory leads to the same quantum mechanics.

Historically, the duality was first recognized for light. It had already emerged in the course of EINSTEIN's work on the statistical dispersion of the radiation field (Chapter 3). As we have seen, the quantum theory of matter was built up from particle mechanics; it developed in this way almost up to the emergence of the SCHROEDINGER equation. Why is it that the matter wave was not seen until so late—1923? The state of experimental technology would hardly have permitted its earlier discovery. But what of the analogy with light? The analogy might well have been obvious to EINSTEIN or EHRENFEST. However, EINSTEIN was by now of course far too busy with his general theory of relativity, and EHRENFEST was still unhappy about the duality of light.

Towards the Matter Wave

In March 1923 William DUANE[1] had already found a relationship between a material momentum p and a kind of wave number $1/l$: if an X-ray of wavelength λ was diffracted by matter with a

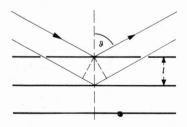

FIGURE 15: DUANE'S FORMULA

periodic structure, the period being l, according to the wave model (Figure 15) the BRAGG condition

$$n\lambda = 2l\cos\vartheta$$

held. The momentum transferred to the matter according to the light quantum hypothesis was

$$\Delta p = 2\frac{h\nu}{c}\cos\vartheta = 2\frac{h}{\lambda}\cos\vartheta.$$

Quantum and wave pictures combine to give

$$\Delta p = n\frac{h}{l}.$$

Thus only certain definite values of the momentum, multiples of h/l, are accepted by matter of period l. In October 1923 COMPTON recognized in this the equation $\oint p\ \mathrm{d}x = hn$ for the quasi-periodic motion of matter. DUANE's equation is formally equivalent to the later DE BROGLIE relationship between the momentum of a particle and the wave number of a wave. But the corresponding concept is not to be found in DUANE's work. If DUANE had applied his knowledge that periodically constructed matter could absorb only discrete momenta to impinging particles of matter, he would have been able to associate a wave with the stream of particles (JORDAN was to indicate this in April 1926[2]). Altogether, no theory which included the duality of light without the corresponding duality of matter would have been consistent.

The concept of characteristic vibrations of an atom is given in a vague form by Marcel BRILLOUIN. [3] He tried to explain the quantum phenomena of the atom in terms of a particular state of the aether in the region of the atomic nucleus, with a diminished propagation velocity. The interior region then had characteristic modes of vibration whose frequencies were of the same order of magnitude as the spectral frequencies, assuming the value of the atomic radius and for suitable velocity of propagation. DE BROGLIE was aware of this work.

The de Broglie Wave

Louis DE BROGLIE had been pondering over light quanta and had tried to reconcile them with light waves. In September 1923 he published the basic ideas of his wave concept of matter.[4] He ascribed a frequency of oscillation to each mass. A particle at rest, of mass m, was given a frequency v_0, which was determined from the equation:

$$mc^2 = hv_0.$$

Using a LORENTZ transformation

$$\bar{t} = \frac{t - vx/c^2}{\sqrt{1 - v^2/c^2}}$$

for a particle moving with velocity v he obtained a plane phase wave

$$\sin 2\pi v\left(t - \frac{v}{c^2}x\right)$$

with phase velocity $u = c^2/v$, frequency $v = v_0/\sqrt{1 - v^2/c^2}$ and wave number

$$\frac{1}{\lambda} = \frac{1}{c^2}\frac{v_0 v}{\sqrt{1-v^2/c^2}} = \frac{mv}{h\sqrt{1-v^2/c^2}}.$$

This wave model helped DE BROGLIE to explain after a fashion the quantum condition for the atom. A phase wave orbiting about the nucleus comprises a whole number of wavelengths:

$$2\pi r = n\lambda.$$

Taking a momentum $p = h/\lambda$, this thus led to the relationship between the angular momentum and the azimuthal quantum number

$$P = mrp = n\frac{h}{2\pi}$$

for the angular momentum of the orbiting particle.

A little later DE BROGLIE epitomized his model thus: the relationship between the new mechanics and classical mechanics is similar to that between wave optics and geometrical optics. He was later (July 1924[5]) to state that the basic equation of his theory was the relationship between the four-vector of energy and momentum (E,p) of a particle and the four-vector of frequency and wave number $(v,1/\lambda)$ of a wave:

$$E = hv \qquad p = \frac{h}{\lambda}. \tag{1}$$

He was even able to explain the combination principle for spectra in terms of his model.[6] If the frequencies v_n and v_{n+1} appear in the wave field, then the beat frequency $v_{n+1} - v_n$ can be observed. He was able to go on to give, for a matter-wave in an electric field with potential U, the equation[7]:

$$hv = \frac{mc^2}{\sqrt{1-v^2/c^2}} + eU. \tag{2}$$

He considered the explanation of the quantum condition to be a matter of prime importance. In his summary of November 1924 he called it the '*première explanation physiquement plausible*'.[7] The hope expressed there—of being able to understand quantum phenomena in intuitive terms—could not, however, be fulfilled.

Duality for Light and Matter

EINSTEIN's research into the statistical fluctuations of energy and momentum in the radiation field, in the course of which he deduced the contributions from light particles and light waves from PLANCK's radiation formula, laid the foundations of the idea of the duality of light (Chapter 3). But others were hardly aware of its significance. Thus problems connected with the radiation formula were once again being discussed by 1923, by, for example, Wolfgang PAULI and Walter BOTHE.[8] Great excitement followed a short notice published by the Indian S. N. BOSE, translated by EINSTEIN.[9] 'As the derivations of the radiation formula given up to now have been unsatisfactory from the logical point of view', he derived it from the light quantum hypothesis with the aid of statistics. The light quanta were distributed among cells of phase space with magnitude h^3. An event was defined by the number of cells with a particular number of quanta of a particular kind (frequency). The events thus corresponded to the occupation numbers of the cells. The BOSE statistics of light quanta was thus the same as that earlier applied by PLANCK for energy quanta (Chapter 2) and thus led to the PLANCK radiation formula. This method of counting events for indistinguishable particles, which had already been perfectly clearly recognized by NATANSON in 1911, was subsequently to be called BOSE statistics (NATANSON's work had of course been forgotten by 1924). It was not until some time later that the alternative possibility of the quantum statistics of indistinguishable particles, that of FERMI statistics, was considered.

EINSTEIN now completed the corresponding modification of the statistics of matter particles.[10] The theoretical behaviour of an ideal gas at low temperatures contradicted the NERNST heat theorem. But by applying BOSE statistics this contradiction could be eliminated. According to EINSTEIN, deviations from the classical ideal behaviour arose when the quantity $nh^3/(mkT)^{3/2}$ became noticeable, where n denoted the number of particles per unit volume, m their mass. At room temperature, n would then have to be of the same order of magnitude as in fluids; at temperatures of a few degrees Kelvin one-thousandth of that amount would suffice. In January 1925 EINSTEIN showed that a gas whose particles satisfied BOSE statistics displayed vibrations that indicated contributions from both particles and waves, and he mentioned DE

BROGLIE's wave model. For many, this was the first that they had heard of it.

Experimental Proof of the Matter Wave

At the time when DE BROGLIE introduced his idea of the matter wave several phenomena were known that were closely connected with the wave nature of matter. These alone, however, would not have been sufficient to lead to the deduction of its wave nature. In July 1925 Walter ELSASSER in Göttingen recognized two phenomena as an expression of these wave properties: the 'RAMSAUER effect', i.e., the diminution of the average action of the atoms of a noble gas with respect to slow electrons when their velocity decreased, and an angular variation, rather like interference, of the intensity for the reflection of slow electrons by the surfaces of certain metals. This had been discovered by C. J. DAVISSON and C. H. KUNSMAN.[11] ELSASSER was helped here by discussions with FRANCK. It was possible to explain the RAMSAUER effect qualitatively in terms of the diffraction of long waves by small spheres, and it was possible that the phenomena that arose at the surfaces of metals could be real interference between waves in the crystal lattice: the latter appeared to fit the DE BROGLIE equation $\lambda = h/mv$ at least in terms of their order of magnitude. Of course, attempts were made at the time to obtain more precise confirmation. According to the DE BROGLIE equation wavelengths of the order of magnitude of the usual X-ray wavelength corresponded to the electron velocities; and these were easy to experiment with. But electrons did not penetrate very far into matter. Thus it was at first more difficult to establish sharp interferences than it was for X-rays. However, in 1927 DAVISSON and L. H. GERNER obtained clear interference maxima for the reflection of electrons by single nickel crystals. In the same year G. P. THOMSON was able to carry out an accurate test of the equation $\lambda = h/mv$ governing interference for the transition of electrons through thin metal sheets, and it was confirmed. In 1930 MARK and WIERL were able to produce interferences between electrons for gas molecules, which were very similar to the X-ray interferences produced by DEBYE.[12]

Indications of interference were also found in atomic radiation, by STERN in 1929 for He rays and crystals of rock salt, and even more clearly by ESTERMANN and STERN using He and H_2 rays and LiF crystals in 1929.[13]

Towards the Schroedinger Equation

It would have been possible to derive the SCHROEDINGER equation in two ways: from the matter wave, or from quantum mechanics. For the motion of homogeneous matter in an electric field the equation

$$\frac{p^2}{2m} + eU = E$$

held, according to the particle model. For the corresponding wave, which was assigned a wave-number vector k (per 2π units of length) that varied with position according to the equation $p = \hbar k$, DE BROGLIE gave (cf. (2)):

$$\frac{\hbar^2}{2m} k^2 + eU = E = \hbar\omega. \tag{3}$$

The quantity k is of course defined exactly only for constant U. The equation thus explains how k changes if a matter wave of fixed frequency goes from a field-free region through a field into another field-free region. In the field itself we must use a wave equation, and the simplest such equation which gives equation (3) with constant U, for a plane wave

$$\psi \sim e^{-i\omega t + ikx}$$

is

$$-\frac{\hbar^2}{2m} \Delta\psi + eU\psi - i\hbar\dot{\psi} = 0 \tag{4}$$

for constant $E = \hbar\omega$:

$$-\frac{\hbar^2}{2m} \Delta\psi + (eU - E)\psi = 0. \tag{5}$$

It is thus possible to obtain (3) only from (5), i.e., for constant E, if we take the real wave

$$\psi \sim \sin(-\omega t + kx + \alpha).$$

As the electric potential U must also comprise the action of the matter upon itself, equations (4) and (5) must be extended by putting

$$\Delta U = -\rho$$

where the density ρ of electric charge must be expressed in terms of ψ. As ρ should depends on ψ^2 or $\psi^*\psi$, it is possible to regard U, in the limiting case of very thinly distributed matters, as the potential of the external field neglecting the reaction of the matter upon itself. (4) and (5) may thus also be written as:

$$-\frac{\hbar^2}{2m}\Delta\psi + (V-E)\psi = 0 \tag{6}$$

where V is the potential energy of a particle in the external field. Such an equation, containing both the particle quantities m and e (which appears in $V = eU$) and the field quantity ψ, is of course illogical. But if we consider the fact that there are particles as well as waves and recall that an individual particle does not act upon itself, it would be possible to take equation (6) as the basis of the theory of a single particle, treating V merely as the potential of the external field. One could express the unit of matter, a particle, by means of

$$\int \psi^*\psi \, d\tau = 1.$$

The alternative route to the SCHROEDINGER equation begins with quantum mechanics, where we calculate in terms of mathematical models, matrices or operators which satisfy commutation relations such as

$$i(pq - qp) = \hbar. \tag{7}$$

Because of the relationship

$$\left(\frac{d}{dq}q - q\frac{d}{dq}\right)F(q) = F(q)$$

which holds for the differential operator d/dq, it is possible to satisfy (7) by replacing p by $\hbar\, d/idq$. More generally it is possible to satisfy the commutation relations of quantum mechanics for several degrees of freedom by using

$$p_k = \frac{\hbar}{i}\frac{\partial}{\partial q_k}. \tag{8}$$

It is thus necessary to replace the classical equation

$$H(p_1, p_2 \ldots q_1, q_2 \ldots) - E = 0$$

by the equation

$$H\left(\frac{\hbar}{i}\frac{\partial}{\partial q_1}, \frac{\hbar}{i}\frac{\partial}{\partial q_2} \ldots q_1, q_2 \ldots\right) \psi(q_1, q_2 \ldots) - E\psi(q_1, q_2 \ldots) = 0$$

where the Hamiltonian function H has now become a Hamiltonian operator. Thus for a single particle, with

$$H = \frac{p^2}{2m} + V(x)$$

we obtain equation (6).

Schroedinger's 1926 Papers

History chose the path via the matter wave. In 1926 Erwin SCHROEDINGER published six papers.[14] The first was completed in January. It gave the prediction of the energy values of the hydrogen atom and tended to conceal the connection with DE BROGLIE's ideas. The second, dating from February 1926, made use of DE BROGLIE's view that the new mechanics was related to classical mechanics as wave optics was to geometrical optics, giving a number of simple applications. If we summarize both papers we see that SCHROEDINGER, stimulated by DE BROGLIE, took an analogy between the mechanics of particles in a field of force and geometric optics in an inhomogeneous medium and expanded it into a wave theory. The analogy had already been recognized by HAMILTON. The quantity $-Et + S(x,y,z)$ which arises in connection with the HAMILTON-JACOBI differential equation in mechanics, where $p = \text{grad } S$, corresponds in optics to the propagation of a wave front. To within a constant factor it is the phase of the wave. SCHROEDINGER now wrote the DE BROGLIE phase wave in the form:

$$\sin \frac{-Et + S(x,y,z)}{\hbar}$$

it had the phase velocity:

$$u = \frac{E}{|\text{grad } S|}.$$

Expressed in terms of mechanical quantities, this would be:

$$u = \frac{E}{p} = \frac{E}{\sqrt{2m(E-V)}}. \qquad (9)$$

Expanding this to a wave theory now meant that a wave equation was needed. It seemed reasonable to put:

$$-\Delta\psi + \frac{1}{u^2}\ddot{\psi} = 0$$

and thus for a homogeneous wave of frequency $\omega = E/\hbar$

$$-\Delta\psi - \frac{E^2}{\hbar^2 u^2}\psi = 0$$

and using (9):

$$-\frac{\hbar^2}{2m}\Delta\psi + (V-E)\psi = 0. \qquad (10)$$

SCHROEDINGER regarded this equation as the basis of the wave theory of particles. With certain physically meaningful conditions on the function ψ, the equation can be solved only for certain values of E, the eigenvalues of the equation. The discrete states, which had up to now required additional conditions, were thus given an almost intuitively plausible explanation, as the characteristic waves. The frequencies of radiation $\omega_n - \omega_l$ that had previously been ascribed to the transition between two states (n,l) were beat frequencies in SCHROEDINGER's theory.

Applying this theory to the harmonic oscillator, SCHROEDINGER obtained the eigenvalues

$$E = \hbar\omega(n + \tfrac{1}{2}).$$

In the case of the multi-dimensional isotropic harmonic oscillator, several eigenfunctions were associated with each energy $E(n)$. SCHROEDINGER also dealt with the rigid rotator with a fixed axis, with a free rotator, the latter leading to

$$E = \frac{\hbar^2 n(n+1)}{2I}$$

and with the biatomic vibrating and rotating molecule. The H atom was treated in the first paper.

Superimposing eigenvalues of different frequencies:

$$\psi = \Sigma c_n u_n(x) e^{-i\omega_n t}$$

where x represents all the position co-ordinates and u_n satisfies the wave equation (10) for a fixed eigenvalue E_n, SCHROEDINGER was able to give a model of concentrated distributions of matter by a suitable choice of the coefficients c_n. For the harmonic oscillator these groups of waves carried out periodic motion and did not diverge. They thus behave like particles. That they did not decompose was in any case soon seen to be a property peculiar to the harmonic oscillator.

In March 1926 SCHROEDINGER was able to demonstrate the formal equivalence between his wave theory and the HEISENBERG-BORN-JORDAN version of quantum mechanics. The basis of the equivalence was of course the equation

$$\left(\frac{d}{dq}q - q\frac{d}{dq}\right)F(q) = F(q).$$

Using a complete, normalized orthogonal system $u_n(x)$, where x represents all the variables, the matrix

$$\Gamma_{mn} = \int u_m(x)\cdot\Gamma u_n(x)\cdot dx$$

could be assigned to an operator Γ. If the operators

$$p_k = \frac{\hbar}{i}\frac{\partial}{\partial q_k}$$

were introduced, precisely the same relationship as given by HEISENBERG, BORN and JORDAN was obtained for the matrices

$$(p_k)_{mn} = \int u_m(x)\frac{\hbar}{i}\frac{\partial u_n(x)}{\partial q_k}dx$$

$$(q_k)_{mn} = \int u_m(x)q_k u_n(x)\,dx.$$

Applying quantum mechanics to a given mechanical system now meant solving a partial differential equation, and in particular calculating its eigenvalues and eigenfunctions.

A paper of May 1926, SCHROEDINGER's 'third communication', adapted perturbation methods to the wave equation, using the substitution

$$\psi = \psi^{(0)} + \lambda \psi^{(1)} + \lambda^2 \psi^{(2)} + \cdots$$

to solve the equation by an iterative procedure. This was a generalization of the procedure that had already been used by RAYLEIGH for the vibrations of a string. SCHROEDINGER was thus able to predict the STARK effect for the H atom together with selection rules and intensities. The fourth paper, dating from June 1926, introduced the time-dependent wave equation. One possibility was to eliminate E from (10) by means of the equation

$$\frac{\partial^2 \psi}{\partial t^2} = -\omega^2 \psi = -\frac{E^2}{\hbar^2}\psi$$

which held for periodic solutions. The equation

$$\left[\left(-\frac{\hbar^2}{2m}\Delta + V\right)^2 \psi - \hbar^2 \ddot{\psi} = 0\right]$$

to which this led did not, however, satisfy SCHROEDINGER. So he now wrote a periodic wave in the complex form

$$\psi \sim e^{\mp i\omega t}.$$

This meant that

$$\dot{\psi} = \mp i\omega\psi = \mp i\frac{E}{\hbar}\psi$$

and the elimination of E from (10) gave:

$$-\frac{\hbar^2}{2m}\Delta\psi + V\psi \mp i\hbar\dot{\psi} = 0. \tag{11}$$

SCHROEDINGER also took this as the basic equation for waves with a variable frequency ω and thus for states with a variable E. He was thus able to treat systems with time-dependent potential energy and to formulate a theory of scattering. He went on in the same paper to give the relativistic generalization of the wave equation which allowed for a magnetic field. Corresponding to the equation

$$-\left(\frac{E}{c} - eU\right)^2 + (\mathbf{p} - e\mathbf{A})^2 + m^2 c^2 = 0$$

for a particle in an electromagnetic field he used the operators
$E \rightarrow i\hbar\partial/\partial t, p_x \rightarrow \pm\hbar\partial/i\partial x,...$ to rewrite the wave equation in the
form:

$$\left\{ -\left(\pm\frac{i\hbar\partial}{c\partial t} - eU \right)^2 + \left(\pm\frac{\hbar\partial}{i\partial x} - eA_x \right)^2 + \cdots + m^2c^2 \right\} \psi = 0. \quad (12)$$

SCHROEDINGER had extended the application of his wave equations (10) and (11) to the case of more than one particle. ψ then became a function in the multi-dimensional space of the co-ordinates. SCHROEDINGER could still say very little as to the meaning of the field quantity ψ. $\psi^*\psi$ represented a weight function in co-ordinate space, which appeared in place of a point in classical mechanics. Equation (11) and its conjugate gave a conservation law for the weight function. This amounted to

$$\frac{\partial}{\partial t}\psi^*\psi + \mathrm{div}\frac{\hbar}{2im}(\psi^* \,\mathrm{grad}\,\psi - \psi\,\mathrm{grad}\,\psi^*) = 0 \quad (13)$$

(we have chosen the negative sign for i in (11)). For larger numbers of particles the corresponding representation can be obtained in multi-dimensional co-ordinate space. The charge and current densities in (13) contain the frequencies $\omega_n - \omega_l$ if the frequencies ω_n and ω_l appear in ψ.

[1] W. DUANE, Proc. Nat. Akad. Sci. **9**, 158 (1923)
[2] P. JORDAN, Z. Phys. **37**, 376 (1926)
[3] M. BRILLOUIN, Compt. Rend. **168**, 1318, **169**, 48 (1919), **171**, 1000 (1920), In. de phys. **3**, 65 (1922)
[4] L. DE BROGLIE, Compt. Rend. **177**, 507, 548 (1923), Phil. Mag. **47**, 446 (1924)
[5] L. DE BROGLIE, Compt. Rend. **179**, 39 (1924)
[6] L. DE BROGLIE, Compt. Rend. **179**, 676 (1924)
[7] L. DE BROGLIE, Ann. de phys. **3**, 22 (1924)
[8] W. PAULI, Z. Phys. **18**, 272 (1923)
   W. BOTHE, Z. Phys. **20**, 145 (1923), **23**, 214 (1924)
[9] S. N. BOSE, Z. Phys. **26**, 178 (1924)
[10] A. EINSTEIN, Sitz. Ber. Berlin **1924**, 261, **1925**, 3, 18
[11] W. ELSASSER, Naturwiss. **13**, 711 (1925)
[12] C. J. DAVISSON and L. H. GERMER, Phys. Rev. **30**, 705 (1927), Proc. Nat. Akad. Sci. **14**, 317 (1928)
   G. P. THOMSON, Proc. Roy. Soc. **117**, 600, **119**, 651 (1928)
   H. MARK and R. WIERL, Z. Phys. **60**, 741 (1930)
[13] O. STERN, Naturwiss. **17**, 391 (1929)
   I. ESTERMANN and O. STERN, Z. Phys. **61**, 95 (1930)
[14] E. SCHROEDINGER, Ann. Phys. **79**, 361, 489, 734, **80**, 437, **81**, 109 (1926), Naturwiss. **14**, 664 (1926)

12. THE COMPLETION OF QUANTUM MECHANICS

Four Versions

BY the spring of 1926 there were four apparently equivalent versions of quantum mechanics. The matrix formulation based on HEISENBERG's work, DIRAC's mechanics of q-numbers, BORN and WIENER's preliminary attempts at a calculus of operators and SCHROEDINGER's wave equation. In classical mechanics a mechanical system is characterized by a Hamiltonian $H(p,q)$ or by equations of motion. The matrix version still used the classical H, but for the variables p_k and q_k matrices were introduced whose rows and columns were actually intended to correspond to the numbers of discrete energy states or even to the values of an angular momentum. $H(p,q)$ must therefore become a diagonal matrix. If the matrix was not yet in this form it was necessary to achieve this by introducing new matrices, using a canonical transformation SqS^{-1}, SpS^{-1} (Chapter 10). The correspondence with classical mechanics was assured by the use of the Hamiltonian, and the commutation relations gave quantum conditions in the limit. DIRAC's mechanics of q-numbers freed itself from the specialized matrix form by treating as fundamental only the algebraic relationships between the p_k and q_k and the quantities which depended on them. The analogy with classical methods was thus still somewhat closer than in the HEISENBERG-BORN-JORDAN formulation as even the quantities $i(pq - qp)$ corresponded asymptotically to classical quantities. The use of operators instead of matrices or q-numbers had not yet been particularly successful. It was hoped to use them to deal with aperiodic phenomena, and thus with those with continuous values of the energy. Finally, the SCHROEDINGER equation, by basing the theory on a trusted field of mathematical analysis, opened an easy route to the solution of quantum-mechanical problems.

But did the wave-like quantity that appeared in SCHROEDINGER's version have any physical reality? And furthermore, was there a general representation of quantum mechanics which included the known forms as special cases? This question was raised because people were by now convinced of the complete equivalence of the various versions of quantum mechanics. Considerable progress was made towards answering both these questions in the same year, 1926.

With hindsight it is possible to make a neat division between the completion of the formalism of quantum mechanics and its physical interpretation. In the actual course of history they were, however, interwoven. The completion of the formal theory by the introduction of 'transformation theory' was to some extent dependent on the probability interpretation. Transformation theory was to explain the relationship between the quantum and classical pictures by means of 'uncertainty'. We shall therefore have to consider the probability interpretation, transformation theory and uncertainty, one by one.

The Probabilistic Interpretation

DE BROGLIE and at first SCHROEDINGER thought that they had made it possible to obtain an intuitive understanding of quantum phenomena. SCHROEDINGER thought so because he had recognized the stationary states of the old quantum theory as characteristic waves in a continuum, and because he thought he could represent particles as tight bundles of waves. But that was an intuitive representation only if the wave function ψ was a function in three-dimensional x, y, z-space for the single particle. In his fourth paper (June 1926) $\psi^*\psi$ indeed denoted a kind of distribution in co-ordinate space which had replaced the well-defined point of classical mechanics.

In a short note in June and a detailed paper in July 1926 BORN gave a tenable physical interpretation of ψ.[1] He saw that of all the versions of quantum mechanics only that of SCHROEDINGER was capable of giving a simple description of aperiodic phenomena. He considered the collision of electrons with atoms. BORN's interest may well have been determined by the fact that his closest colleague FRANCK had worked on electron collision and had given it a great deal of thought. BORN saw the DE BROGLIE or SCHROE-DINGER wave as a kind of conducting field for electrons. The field itself was determined causally and the magnitude of the field in turn determined the probabilities for the transition of an atom into other states and for the deflection of an electron in particular directions. BORN first tackled the scattering of an electron by a spherically symmetrical potential $V(r)$. As the wave functions could be regarded as periodic in time he needed only the time-independent SCHROEDINGER equation

$$-\frac{\hbar^2}{2m}\Delta\psi + [V(r) - E]\psi = 0.$$

He discovered solutions which contained one ingoing plane wave

$$\psi \sim e^{ikz - i\omega t}$$

and one outgoing scattering wave

$$\psi = f(k,\vartheta)\frac{e^{ikr - i\omega t}}{r}.$$

He regarded the quantity

$$|f(k,\vartheta)|^2\,d\Omega$$

for suitable normalization of the incoming wave as the probability that an electron would be deflected through an angle ϑ into the solid angle $d\Omega$. $\psi^*\psi\,d\tau$ was then the probability that the co-ordinates would lie in the region $d\tau$. The second exercise, that of the scattering by an atom with stationary states n, gave amounts

$$|f_{nm}(\vartheta)|^2\,d\Omega$$

for the probabilities of deflection into the solid angle $d\Omega$ in the course of the transition from the state n into the state m.

The concept of the transition to another stationary state had not really been understood in the philosophical context of the BOHR theory or that of HEISENBERG, BORN and JORDAN. In the SCHROEDINGER formulation it seemed that the concept of a mixture of states in vibration appeared in its place in SCHROEDINGER's theory. In October 1926 BORN considered phenomena of the type

$$\psi = \Sigma c_n u_n(x)e^{-i\omega_n t}.$$

These are solutions of the time-dependent SCHROEDINGER equation

$$-\frac{\hbar^2}{2m}\Delta\psi + V(x)\psi - i\hbar\dot{\psi} = 0$$

if the functions $u_n(x)$ are the solutions of the time-independent SCHROEDINGER equation for the energy $E_n = \hbar\omega_n$. BORN now regarded $|c_n|^2$ as the probability that the system was in the stationary state n, which explained the SCHROEDINGER theory 'in HEISENBERG terms'. Under an external action given by the addition of $W(x,t)$ to $V(x)$, starting at time $t = 0$, the initial $\psi_n = u_n(x)\,e^{-i\omega_n t}$ became a function

$$\psi_n(x,t) = \sum_m f_{nm}(t)u_m(x)\,e^{-i\omega_m t}.$$

Once the external influence had been removed ($W = 0$ for $t > T > 0$) he thus obtained

$$\psi_n(x,t) = \sum_m b_m u_m(x)\,e^{-i\omega_m t}.$$

The quantities $E_m = \hbar\omega_m$ once again denoted the energies and the u_m the eigenfunctions of the possible stationary states. The transition probabilities $|b_{nm}|^2$ were determined by u_n and by the nature of the external influence. For an arbitrary initial state $\sum c_n u_n(x)\,e^{-i\omega_n t}$ they were given by the state and by the external influence. Under slow change of $W(x,t)$ no transitions occurred. Thus the adiabatic law had been proved in the new quantum mechanics.

BORN's interpretation of the SCHROEDINGER function ψ was very quickly generalized. PAULI referred in a letter to the *probability* $|\varphi(p)|^2\,dp$ for a momentum value p in the region dp.[3] In December 1926 he explained $|u_n(x)|^2\,dx$ as the probability that the system was in the state n and that the co-ordinates represented by x lay in the interval dx.[4] More generally he expected to find a function $\varphi(\beta,x)$ that would express the probability that the position x lay in dx as $|\varphi(\beta,x)|^2\,dx$, for a constant value β of some other quantity. φ was later to be called the probability amplitude. And here lay the realization that it was possible to fix the value of one quantity and to search for the probability of a particular value of some other quantity.

Transformation Theory

A brand new version of quantum mechanics, breaking out of all its historical shackles, was now very much in the air. The matrix version established the numbers n, which corresponded to states with definite values of the energy and perhaps also of angular momentum. The SCHROEDINGER formulation emphasized the position x in co-ordinate space; the time-independent version also established the value of n and gave the probability of the position x.

In the search for a generalized version of quantum mechanics the most important factor was the recognition of the SCHROE-DINGER equation as the method of carrying out a canonical transformation, which transforms $H(p_1, ..., q_1, ...)$ into a diagonal matrix. The first steps were a more precise investigation of the canonical transformations themselves.[5] In April 1926 JORDAN proved that all canonical transformations are of the form $\bar{A} = SAS^{-1}$, which becomes $\bar{A} = SAS^*$ for real quantities. Fritz LONDON in May and JORDAN in July saw that the classical canonical transformations could be preserved. Using series expansions of $\psi(x) = \sum \psi_n u_n(x)$ in terms of orthogonal sets of functions $u_n(x)$ LONDON assigned matrices to the differential operators, e.g., the $\hbar \partial / i \partial q$ which appeared in the SCHROEDINGER equation. He discovered that in general if an operator is transformed according to the formula

$$\bar{A} = SAS^* \qquad \bar{A}_{ik} = S_{in} A_{nm} S^*_{mk}$$

this corresponds to the transition to another orthogonal system and to a transformation

$$\bar{\psi} = S\psi \qquad \bar{\psi}_n = S_{nm} \psi_m$$

of the series coefficients (September 1926). A transformation of the coefficients of this kind denotes a rotation in the Hilbert space of the functions. The coefficients ψ_n play the role of vector components (see the Appendix). The canonical transformations were therefore co-ordinate transformations in an abstract space, and the matrices which corresponded to the physical quantities were representative matrices.

Up to now consideration had mostly been given to matrices A_{nm} or systems of coefficients ψ_n with discrete suffices. The extension of the transformations to systems with continuous suffices was achieved by the 'transformation theory' which was discovered independently by DIRAC and JORDAN in December 1926.[6] They started out from different points. DIRAC tried to adjust the formalism of his q-numbers to the SCHROEDINGER equation. JORDAN worked from PAULI's idea of the probability $|\varphi(\beta, q)|^2 \, dq$. DIRAC replaced the transformation

$$\bar{A}_{\xi' \xi''} = S_{\xi' \alpha'} A_{\alpha' \alpha''} S^*_{\alpha'' \xi''}$$

for continuous suffices by

$$\bar{A}(\xi'\,\xi'') = \int S(\xi'\,\alpha')\,\mathrm{d}\alpha'\,A(\alpha'\,\alpha'')\,\mathrm{d}\alpha''\,S^*(\alpha''\,\xi'')$$

where dashes denote eigenvalues. In a diagonal matrix ξ with the eigenvalues ξ', ξ'' he replaced the expression

$$\xi(\xi'\,\xi'') = \xi'\,\delta_{\xi'\xi''}$$

by

$$\xi(\xi'\,\xi'') = \xi'\,\delta(\xi' - \xi'')$$

using the (DIRAC) delta-function, defined by

$$\int f(x)\delta(x-y)\,\mathrm{d}x = f(y).$$

JORDAN laid down axioms for the 'probability amplitude' $\varphi(\beta,q)$, its independence of the dynamic behaviour, the superposition of amplitudes, and symmetry with respect to β and q. Both writers investigated the equations for canonical conjugate variables p and q, obtained the same results and finally showed that the SCHROEDINGER eigenfunctions $u_{n'}(q')$ were the elements $S(n'q')$ of the transformation matrix S, which diagonalizes the Hamiltonian operator. This transformation theory developed by LONDON, DIRAC and JORDAN was unable to satisfy the rigorous requirements of mathematics. Its mathematical development was carried out as a result of a lecture on the latest developments in quantum mechanics given by HILBERT in the winter term 1926–7 in Göttingen. In preparing his lecture he was assisted by NORDHEIM and (in respect of its mathematical content) by John VON NEUMANN.[7] A paper published by the three authors in April 1927 made the DIRAC and JORDAN ideas much clearer and more comprehensible. It managed to make a sharper distinction between the formalism and its physical interpretation. In the next few years VON NEUMANN filled in the gaps in the mathematical argument.[8] It was very difficult indeed to prove the isomorphism of the space of quadratically integrable functions $\psi(x)$, with a finite value of $\int \psi^*\psi\,\mathrm{d}x$, and the space of the coefficients ψ_n with a finite solution of $\sum \psi_n^*\psi_n$, and to demonstrate conclusively that the solution of the quantum-theoretical problem was in fact a transformation of the major axis in these spaces.

Is it possible to regard the establishment of transformation theory as a completion or 'crowning' of quantum mechanics? It bears a closer relationship to the classical particle than to the matter wave. Ideas such as stationary states and quantum jumps had been reinterpreted, but they were still used. From the beginning, when the Hamiltonian was first used in quantum mechanics, the number of particles in a mechanical system had been regarded as being definitely fixed. This restriction was soon relaxed. The wave or field picture was thus again strengthened. But we shall understand this better once we have discussed the connection between the statistics of indistinguishable particles and the symmetry of the wave function (Chapter 13). We may regard the proof of the equivalence of the particle and wave aspects (Chapter 15) as the true crowning of quantum theory.

Uncertainty in Quantum Theory

The transformation theory that had been developed towards the end of 1926 had shown which questions were meaningful within the scope of quantum mechanics and could thus be answered. It had achieved this in a rather abstract manner. However, simultaneously with its development, the search continued for a deeper physical meaning and an intuitive interpretation of the theory.[3] In a letter to PAULI, HEISENBERG wrote in October 1926 more or less as follows: the fact that there is no sense in speaking of a monochromatic wave in a brief time interval corresponds in the wave formulation to the equation $i(pq - qp) = \hbar$. Analogously it is quite pointless to speak of the position of a corpuscle of a definite velocity. However it would be meaningful if we did not take position and velocity so literally. It was probably around then that PAULI said that we could look at the world with a *p* eye or with a *q* eye, but if we tried to do both we went cross-eyed. Apart from the correspondence between HEISENBERG and PAULI, there were continuing discussions between BOHR and HEISENBERG. The approaches taken by these two men were quite different. BOHR saw the wave-particle dualism as the starting point for the physical interpretation of quantum phenomena; while for HEISENBERG the theory rested on the theorems of quantum mechanics and transformation theory. HEISENBERG was able to clarify his own view around February 1927. A long letter he wrote to PAULI gives roughly the content of HEISENBERG's paper on the uncertainty principle. 'This marks the dawn of quantum theory,' was PAULI's reply. But BOHR was not yet quite satisfied. However, HEISENBERG now published (March 1927) his paper 'On the Intuitive Content of Quantum-theoretical Kinematics and Mechanics'.[9] It gave an analysis of the fundamental concepts of position, velocity, path and energy of a particle.

The position of a particle can be determined by using a microscope. In doing this, one must allow for an uncertainty of the order of magnitude of the wavelength of the light used:

$$\Delta x \approx \lambda.$$

This may be imagined arbitrarily small, so that the position can thus in principle be determined with arbitrary precision. But light of very short waves leads to an observable change in the momentum of the observed particle, and to an uncertainty of

$$\Delta p \approx h/\lambda.$$

in the momentum so that the uncertainty in positions and the uncertainty in the momentum are related to each other by the formula

$$\Delta x \, \Delta p \approx h.$$

A similar argument holds if one determines the position of the particle by its ability to deflect material particles instead of using a light microscope. For a 1s electron, the product of the orders of magnitude of the distance from the nucleus and of the momentum is already equal to roughly h, so that there is no sense in talking of the orbits of a 1s electron. The velocity of a particle can, for example, be determined by means of the Doppler effect for light, but only subject to an uncertainty $\Delta p \approx h/\lambda$ in the momentum, because of the recoil that occurs; for a simultaneous determination of position we have $\Delta x \approx \lambda$. The energy of an atomic state is given by the frequency v transmitted, which can be determined more accurately the longer the time available, giving, in fact,

$$\Delta v \, \Delta t \approx 1.$$

We thus have the relationship

$$\Delta E \, \Delta t \approx h$$

between the uncertainty in the determination of energy and the evaluation of a point in time. Classical concepts can be defined precisely, even in the atomic domain. The simultaneous determination of two canonically conjugate quantities is however subject to an uncertainty given by $\Delta p \, \Delta q \approx h$. If there were experiments that allowed of a more precise determination of the variables then quantum theory would be impossible. (It might have been more accurate for HEISENBERG to say that this would lead to contradictions with established results in quantum theory.)

HEISENBERG used the DIRAC-JORDAN transformation theory to predict the uncertainties. In the simplest case the probability amplitude for the position

$$\psi(x) \sim e^{-\frac{x^2}{2a^2}} \qquad \psi^* \psi \sim e^{-\frac{x^2}{a^2}}$$

corresponded to one for the momentum

$$\varphi(p) \sim e^{-\frac{p^2}{2b^2}} \qquad \varphi^* \varphi \sim e^{-\frac{p^2}{b^2}}$$

where

$$ab = \hbar.$$

HEISENBERG also showed how such wave packets dispersed with time.

HEISENBERG expressed the consequences of uncertainty as follows: *in the assertion 'if we know the present we can predict the future' it is not the deduction but the premise which is false.* As everything that happens is subject to the relationship $\Delta p \, \Delta p \approx h$ the law of causality is thus invalid as stated.

While HEISENBERG derived his uncertainty relationship in the context of the transformation theory, it was duality that was of major importance for BOHR. He lectured on this version at the congress of physicists that was held in Como in September 1927.[10] In the thought-experiment to determine position by means of a microscope he set great store by the use of wave properties which led to an uncertainty

$$\Delta x = \frac{\lambda}{\sin \varphi}$$

in position (φ is the aperture of the resolving lens), and by the use use of particle properties which led to the Compton effect and to an uncertainty

$$\Delta p = \frac{h \sin \varphi}{\lambda}$$

in the x component of momentum (because of the variation in angle). Even in the basic equations

$$E = h\nu \qquad p = \frac{h}{\lambda}$$

particle and wave properties are interwoven. BOHR's whole epistemology is permeated by the concept of the 'complementarity' of quantities for which the precise knowledge of the one excludes the possibility of a precise knowledge of the complementary quantity. Canonically conjugate quantities in mechanics are complementary. The measurement of the one quantity is bound up with the reaction with the measuring apparatus which disturbs the measurement of the second, complementary, quantity.

At the Solvay Congress in October 1927 (see Frontispiece) DE BROGLIE, BORN, HEISENBERG and SCHROEDINGER lectured on the various versions of the explanation of quantum phenomena.[11] For DE BROGLIE the density that corresponded to the wave function had now become a probability. BORN and HEISENBERG emphasized that quantum theory was now complete. Deviations from the explanation given by transformation theory would involve contradictions of the experiments. The connection between the DE BROGLIE wave and the quantity ψ in co-ordinate space was not yet clear to SCHROEDINGER. ψ must somehow ensure that the actual system corresponds to the classical system that is present in all possible states. In the discussion DE BROGLIE and SCHROEDINGER finished up by arguing that the situation was provisional, while BORN and HEISENBERG regarded it as conclusive.

Links with Earlier Work

Attempts were of course also made to clarify the connection between the SCHROEDINGER theory, the provisional results of old quantum theory and classical mechanics, using simple examples.

A technique developed by WENTZEL, KRAMERS and BRILLOUIN, known as the WKB approximation, showed the extent to which the earlier results of the phase integral method had been approximately correct.[12] In June 1926 WENTZEL transformed the one-dimensional SCHROEDINGER equation

$$-\frac{\hbar^2}{2m} \frac{\mathrm{d}^2\psi}{\mathrm{d}x^2} + [V(x) - E]\psi = 0$$

using the substitution

$$\psi = e^{\frac{i}{\hbar}\int y\,dx}$$

into the form:

$$\frac{\hbar}{i} y' + y^2 + 2m(V-E) = 0.$$

In the limit, as $\hbar \to 0$, this led to

$$y = \pm\sqrt{2m(E-V)} = \pm p$$

$$\psi = e^{\frac{i}{\hbar}\int p(x)\,dx}.$$

Thus $p/\hbar$ was a kind of wave number. By expanding y as a series in $\hbar$, WENTZEL obtained improved approximations. In the case where the classical motion took place in the region $V \leq E$ between two limits x_1 and x_2, WENTZEL found, albeit not entirely rigorously, that

$$\int_{x_1}^{x_2} p\,dx = \frac{hn}{2}$$

$$\oint p\,dx = hn.$$

He applied his method also to a separable system with several degrees of freedom. Léon BRILLOUIN produced identical results in July 1926 independently of WENTZEL. In September 1926 H. A. KRAMERS took a closer look at the behaviour of the solution at the points x_1 and x_2. He took the exact solution for linear $V(x)$ and compared its asymptotic behaviour at some distance from x_1 and x_2 with the behaviour of the approximation

$$e^{\pm\frac{i}{\hbar}\int p\,dx} \qquad e^{\pm\frac{i}{\hbar}\int \bar{p}\,dx} \qquad \bar{p} = \sqrt{2m(V-E)}.$$

He found that

$$\psi = \frac{1}{\sqrt[4]{E-V}} \sin\left(\frac{1}{\hbar}\int_{x_1} p\,dx + \frac{\pi}{4}\right)$$

for $x > x_1$, $V < E$, and thus that

$$\oint p\,dx = h(n+\tfrac{1}{2}).$$

The analogy between the behaviour of a wave packet and a particle was demonstrated very simply by EHRENFEST in September 1927.[13] For a solution of the one-dimensional wave equation

$$-\frac{\hbar^2}{2m}\frac{\partial^2\psi}{\partial x^2} + V(x)\psi - i\hbar\dot{\psi} = 0$$

and for the solution $\psi^*(x)$ of the conjugate equation, he derived the equation:

$$m\frac{d^2}{dt^2}\int\psi^*\psi\,dx = \int\psi^*\left(-\frac{dV}{dx}\right)\psi\,dx$$

or briefly

$$m\ddot{\bar{x}} = -\overline{\frac{dV}{dx}}\,.$$

Thus NEWTON's second law holds for the mean values of position and force.

[1] M. BORN, Z. Phys. **37**, 863, **38**, 803 (1926)
[2] M. BORN, Z. Phys. **40**, 167 (1926)
[3] cf. W. HEISENBERG, Pauli Memorial Volume (1960)
[4] W. PAULI, Z. Phys. **41**, 81 (1927)
[5] P. JORDAN, Z. Phys. **37**, 383, 513 (1926)
   F. LONDON, Z. Phys. **37**, 915, **40**, 193 (1926)
[6] P. A. M. DIRAC, Proc. Roy. Soc. **113**, 621 (1927)
   P. JORDAN, Z. Phys. **40**, 809, **41**, 797 (1927)
[7] D HILBERT, J. v. NEUMANN and L. NORDHEIM, Math. Ann. **98**, 1 (1927)
[8] J. v. NEUMANN, Mathematical Foundations of Quantum Mechanics, Princeton N.J. 1955
   P. JORDAN, Z. Phys. **40**, 809, **41**, 797, **44** (1927)
[9] W. HEISENBERG, Z. Phys. **43**, 172 (1927)
[10] N. BOHR, Naturwiss. **16**, 245 (1938)
[11] Rapport 5me conseil de physique à Bruxelles, Paris 1928
[12] G. WENTZEL, Z. Phys. **38**, 518 (1926)
   L. BRILLOUIN, Compt. Rend. **183**, 24 (1926), Jn. de phys. **7**, 353 (1926)
   H. A. KRAMERS, Z. Phys. **39**, 828 (1926)
[13] P. EHRENFEST, Z. Phys. **45**, 455 (1927)

13. THE EXPLOITATION OF SYMMETRY

Symmetry of a System, Symmetry of States

ANY symmetry or invariance of a mechanical system leads to certain symmetrical properties in its states; in quantum mechanics, to some form of symmetrical behaviour of the SCHROEDINGER eigenfunctions. In the simple case of a one-dimensional potential

$$V(x) = V(-x)$$

with 'cosine' (i.e., reflection) symmetry, the SCHROEDINGER equation

$$-\frac{\hbar^2}{2m}\psi'' + (V-E)\psi = 0$$

immediately implies that for every solution $\psi = u(x)$, both $u(-x)$ and $u(x) \pm u(-x)$ are also solutions for the same eigenvalue. It is easy to see that in the one-dimensional case only a single eigenfunction corresponds to each eigenvalue, so that one of the functions $u(x) \pm u(-x)$ must be zero, i.e., u must be an even or odd function of x. The integral $\int \psi_1^* x \psi_2 \, dx$, which governs dipole radiation, is non-zero only if ψ_1 and ψ_2 are of different types.

The symmetry of the Hamiltonian leads to a partition of the states into terms of different symmetrical behaviour of the eigenfunction and to a selection rule. This connection emerged during 1926–7 for the various symmetries and invariances that arose: for identical particles, for reflection and rotational symmetry and for translation invariance. Certainly, the previous assignment of quantum numbers had already often been characterized by a partition into such systems. The first result was the discovery of the special relationships that held in systems of identical particles.

Indistinguishable Particles

A number of the special features of systems of identical particles were already known independently of the SCHROEDINGER equation. In the course of deriving his radiation formula (1900) PLANCK had applied a form of statistics which is worth noting to the distribution of energy quanta among the oscillators: events of equal probability were the occupation numbers of the oscillators. In 1911 NATANSON had recognized this as containing a possible form of the statistics of indistinguishable particles. BOSE applied the same form of statistics (1924) to light particles and EINSTEIN to gas molecules. EINSTEIN showed that the fluctuations in a BOSE gas behaved as though they were caused by both particles and waves.

Against this 'BOSE statistics' for a gas, in Rome in March 1926 Enrico FERMI set up a different form, later to be known as 'FERMI statistics'.[1] A BOSE-EINSTEIN gas did not in fact satisfy the PAULI exclusion principle. In accordance with the PAULI principle, FERMI required each elementary cell h^3 of phase space for a molecule of a monatomic ideal gas to contain at most one molecule. A permutation of indistinguishable particles should not be regarded as leading to a separate event.

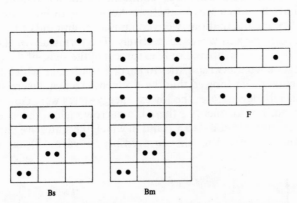

FIGURE 16: BOSE, BOLTZMANN AND FERMI STATISTICS

Figure 16 displays the various possibilities for the distribution of two particles among three boxes according to BOSE, BOLTZMANN and FERMI statistics. For a large number of boxes and particles FERMI obtained the following expression for the number of particles in a macroscopic box, number l:

$$N_l = \frac{1}{e^{\frac{\varepsilon_l - \zeta}{kT}} + 1}. \qquad (1)$$

ε_l denotes the energy of a particle in box l, and ζ is a parameter connected with the total number N of particles, so that we must have

$$\sum_l N_l = N.$$

FERMI discovered a different form for the degeneration of a gas (deviation from the BOLTZMANN gas) from that of EINSTEIN. The deviation was noticeable even for values of $h^3 n/(\mu k T)^{3/2}$ that were no longer small (n is the number of particles per unit volume and μ their mass). The three forms of statistics can be expressed in a single formula:

$$N_l = \frac{1}{e^{\frac{\varepsilon_l - \zeta}{kT}} + \vartheta} \quad (\vartheta = 0, \pm 1).$$

The relationship between N_l and $(\varepsilon_l - \zeta)/kT$ is shown in Figure 17 for FERMI statistics.

PAULI carried the investigation further in December 1926.[2] He regarded both BOSE and FERMI statistics as possible, after HEISENBERG had shown their connection with the symmetry of the SCHROEDINGER wave function. According to both forms of statistics the fluctuations were composed of amounts that came from particles and waves respectively. The formulae for the two types of statistics differed only in respect of the signs of the terms. PAULI applied FERMI statistics to particles with spin, which FERMI had ignored. He showed that the electrons in a metal were almost completely degenerate: the lowest energy states were completely occupied. For an energy ζ there is a narrow region, the extent of which is roughly equal to kT and in which the states are partially occupied. Above this they are empty (Figure 17); at room

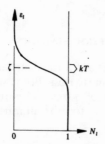

FIGURE 17: FERMI STATISTICS   FIGURE 18: PARAMAGNETISM

temperature $kT \ll \zeta$. In an external magnetic field there is an excess of electrons with one of the two directions of spin.

ε_l includes an additional magnetic energy term that has a different sign for the two directions of spin. This gives a magnetic moment which is proportional to the adduced field and which is not noticeably dependent on temperature. It is thus a magnetic susceptibility, which we may regard as independent of the temperature. In Figure 18 the magnetic energy has not been included in the value of the ordinate: it gives different curves $N(\varepsilon)$ for the two directions of spin. In the course of this work PAULI saw for the first time the fundamental difference between the electron gas for a metal and an ordinary gas. This marked the beginning of a quantum theory of the electronic structure of metals.

L. H. THOMAS gave a statistical method for calculating atomic properties.[3] As it approximated actual circumstances by means of a constant electron density its results were inexact for small numbers of electrons, for atoms whose number nearly completed a shell and at the edge of the atom. THOMAS's paper of November 1926 took FERMI statistics into account by putting an amount of matter that corresponded to two electrons into each cell h^3 of phase space. The momenta p had thus been represented up to a limit of $p^2/2m = eU$, where U is the potential. This was thus given by the equation:

$$\Delta U = \alpha U^{3/2}.$$

FERMI rediscovered the method in February 1928 independently of THOMAS; used it to calculate the atomic number for which p, d and f electrons appeared and thus completed the BOHR theory of the periodic system (Chapter 8). Not long afterwards he calculated the RYDBERG corrections for the s terms.[4]

Symmetry and Statistics

The connection between BOSE and FERMI statistics on the one hand and the symmetry of states of systems of identical particles on the other was demonstrated by HEISENBERG in June 1926 in a paper entitled 'The Many-Body Problem and Resonance'.[5] For two identical coupled oscillators with deflections x_1 and x_2 the combinations $x_1 \pm x_2$ execute independent oscillations. The two frequencies ω_+ and ω_- are different, and the more so the stronger the coupling. A 'resonance splitting' emerges. In quantum theory in an harmonic approximation there are the energies

$$E = \hbar\omega_+ (n_+ + \tfrac{1}{2}) + \hbar\,\omega_-(n_- + \tfrac{1}{2})$$

with the selection rule for dipole radiation $\Delta n_+ = \pm 1$ and $\Delta n_- = 0$; for the radiation of arbitrary multipoles we still have $\Delta n_- = 0, \pm 2, \pm 4, \ldots$. There are thus two systems of states which do not combine, the one with even and the other with odd n_-. Each system of states represents a complete solution. Figure 19 gives the states for weak coupling. The systems which do not combine are denoted by dots and crosses. On the left are the quantum numbers of the uncoupled oscillators. The system with dots corresponds to BOSE statistics, the one with crosses to FERMI statistics.

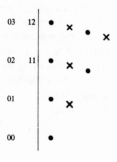

FIGURE 19: TWO IDENTICAL OSCILLATORS

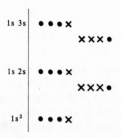

FIGURE 20: THE HELIUM SPECTRUM

First HEISENBERG showed in the language of matrix mechanics that two identical coupled systems always behaved like the two oscillators. Then he stated the SCHROEDINGER functions in the approximation for weak coupling:

$$\psi(1,2) = \frac{1}{\sqrt{2}}[a(1)b(2) \pm b(1)a(2)].$$

The numbers correspond to the two weakly coupled systems (for example, two electrons), and the letters denote eigenfunctions of the uncoupled systems. The functions ψ of the one system are symmetrical in the two co-ordinates 1, 2, i.e., $\psi(1,2) = \psi(2,1)$. Those of the other are antisymmetric, i.e., $\psi(1,2) = -\psi(2,1)$. The fact that the two systems of states would not combine now followed from the symmetry

$$\int [a(1)b(2)+b(1)a(2)]\, f(1,2)[a(1)b(2)-b(1)a(2)]\, d\tau = 0$$

as f had to be symmetrical in both numbers.

The two electrons of the helium atom are identical coupled systems of this kind. Neglecting spin, HEISENBERG recognized the two incompatible systems of terms as the para- and ortho-term systems. When spin was added, weak combinations between the two systems were allowed and the terms split into four. The new terms decomposed once again into two incompatible systems as in Figure 20. Only the one denoted by x actually exists in nature. More precise investigation of atoms with two electrons followed in July using the SCHROEDINGER approach. Two spins can be related in three ways symmetrically and in one way antisymmetrically. If α and β denote the two spin states the symmetric combinations are

$$\alpha(1)\alpha(2) \qquad \alpha(1)\beta(2)+\beta(1)\alpha(2) \qquad \beta(1)\beta(2)$$

and the antisymmetric one is

$$\alpha(1)\beta(2)-\beta(1)\alpha(2).$$

States with a symmetrical orbit eigenfunction (para-terms) can be made antisymmetric in one way by adding spin, terms with an antisymmetric eigenfunction (ortho-terms) in three ways. In fact the para-system consists of singlets and the ortho-system of triplets. So HEISENBERG deduced that in nature only states with antisymmetric eigenfunctions in their electron co-ordinates can arise. The eigenfunction is composed of the path and spin contributions.

In August 1926 DIRAC discovered the same results, also by using the SCHROEDINGER equation.[6] A state in which the first electron has a quantum number n and the second m is not an observable state. In place of the states (n,m) and (m,n) there is only one. For its eigenfunction there are two possibilities:

$$\psi(1,2)=a(1)b(2)\pm b(1)a(2).$$

If we admit the symmetric function BOSE statistics hold, while for antisymmetric ones we must use FERMI statistics.

The work was also extended to more than two particles. In June 1926 HEISENBERG considered the interaction of particles in the states a, b, c,... and gave particular emphasis to the composite states, which are given approximately by

$$\psi(1,2,3...)=\Sigma(-1)^{P}a(1)b(2)c(3)... \tag{2}$$

The summation is to be carried out over all permutations of the particle numbers 1, 2, 3,..., and P is the number of transpositions required to achieve each permutation. These functions, which did not combine with functions of different symmetry behaviour, were antisymmetrical in any two particles and obeyed the PAULI principle. DIRAC showed in August 1926 that the symmetrical combinations

$$\psi \sim \Sigma a(1)b(2)c(3)... \tag{3}$$

corresponded to BOSE statistics, while the antisymmetrical ones (2) corresponded to FERMI statistics.

In December 1926 HEISENBERG considered molecules with two identical nuclei. The eigenfunctions of the electron motion is symmetrical or antisymmetrical in both nuclei, the oscillation adds a symmetrical contribution, increase in rotation gives a symmetrical (antisymmetrical) total function with rotation quantum numbers 0, 2, 4,... for a symmetric (antisymmetric) contribution from the electrons and with the rotation quantum numbers 1, 3, 5,... for an antisymmetrical (symmetrical) electron contribution. It was now possible for HEISENBERG to explain the change of intensity that had been discovered by MECKE in the band spectra of molecules with two identical nuclei. The states whose total function was symmetrical in the identical nuclei and those states with antisymmetric total function formed a pair of incompatible systems. They could arise independently of each other and thus also with different weight. The one type of intensity change (Figure 21b) displayed both systems with different weights, the other type (Figure 21a) showed that only one of the systems appeared. Imposition of a nuclear spin $\hbar/2$ could create circumstances as in the case of ortho- and para-helium, and thus an intensity change of $3:1$.

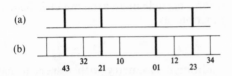

(a)

(b)

43 32 21 10 01 12 23 34

FIGURES 21a AND 21b: TYPES OF SERIES OF ROTATION LINES

The anomalous behaviour of the specific heats of hydrogen at low temperatures was explained by D. M. DENNISON (June 1927) after some preliminary work by HUND.[7] Hydrogen was a mixture of para and ortho hydrogen. The symmetrical rotation states $0, 2, 4 \ldots$ with respect to the two protons of the para share could be antisymmetrized in one way by the imposition of a proton spin $\hbar/2$, while the antisymmetric rotation states $1, 3, 5, \ldots$ of the ortho part could be antisymmetrized in three ways. The statistical weights calculated in this way give the predicted behaviour of the specific heats. Protons thus had a spin $\hbar/2$ and obeyed FERMI statistics.

Group Theory

We have seen from examples that a particular invariance property or—what amounts to the same thing—a particular symmetry in the physical system leads to a partition of the states according to the symmetry behaviour of their eigenfunctions and to selection rules. Thus reflection symmetry,

$$V(x) = V(-x)$$

gave the partitioning into even and odd eigenfunctions, the identity of two particles,

$$H(1,2) = H(2,1)$$

the partitioning into symmetrical and antisymmetrical eigenfunctions with a proscription of combination. Moreover, it was known that a particle in a spherically symmetrical field had eigenfunctions

$$\psi = f(r)\, Y_l(\vartheta,\varphi) \tag{4}$$

where Y_l is a surface spherical harmonic function of the l-th order, and that the selection rule followed for dipole radiation. It was also known that for a particle in an axisymmetric field one could write the eigenfunctions as

$$\psi = f(z,r)\, e^{im\varphi} \qquad \Delta m = 0, \pm 1 \tag{5}$$

in the case of additional symmetry with respect to each plane passing through the axis of rotation as

$$\psi = f(z,r) {\cos \atop \sin} \lambda\varphi \qquad \Delta\lambda = 0, \pm 1. \tag{6}$$

It was possible to deal with the splitting which occurred when the symmetry was decreased, for example as a result of the transition from spherical symmetry to axial symmetry, by writing the spherical function in the form

$$Y(\vartheta,\varphi) = P(\vartheta)\, e^{im\varphi}.$$

Now, the systematic theory of symmetry is simply the mathematical theory of groups, and its introduction into quantum theory enriched the latter with a very powerful tool. In some of the above examples of symmetry the dependence of the eigenfunction or the variable or variables involved in the symmetry was given explicitly. Thus the behaviour of the eigenfunctions and their partitions into systems of different symmetry (like $l = 0, 1, 2,\ldots$; $m = 0, \pm 1, \pm 2,\ldots$; $\lambda = 0, 1, 2,\ldots$) could be seen straight away. This method fails if there is a large number of variables. But in the case of invariance with respect to rotation about an axis or with respect to reflection in any plane through the axis one can, for example, also characterize the behaviour of the functions (6) by means of the transformation matrices

$$\begin{pmatrix} \cos\lambda\alpha & -\sin\lambda\alpha \\ \sin\lambda\alpha & \cos\lambda\alpha \end{pmatrix} \qquad \begin{pmatrix} 1 & 0 \\ 0 & -1 \end{pmatrix} \tag{7}$$

These matrices represent the transformations of the two functions under a rotation $\varphi \to \varphi + \alpha$ and a reflection $\varphi \to -\varphi$ respectively. This method can be taken much further. Transformation matrices of this kind form the 'representations of the group by means of linear transformations', and the 'irreducible' representations of the group of the covering operations of a physical system corations pond to the systems of terms. This form of group theory was first applied to quantum theory by Eugene WIGNER in November 1926.[8] The most important groups in quantum theory are the group of permutations (for identical particles), the reflection and rotation groups and the group of translations.

Permutations

In the first of his two papers of November 1926 WIGNER dealt with a system of three coupled electrons without spin. He gave the six combinations of the functions a, b and c of uncoupled electrons with which the eigenfunctions with coupling are continuously connected. The first two are:

$$\sum a(1)b(2)c(3)$$
$$\sum (-1)^P a(1)b(2)c(3).$$

The summation is to be carried out over all six permutations of the numbers 1, 2 and 3, and P is once again the number of transpositions. Then there were four others of which two each belong to the same eigenvalue. In the second paper he dealt with N coupled electrons using the theory of the permutation group of N elements. This gave the incompatible systems of terms as corresponding to the numerical partitioning of the integers N. In particular for $N = 4$ this gives the decompositions

$1+1+1+1$	(1)
$1+1+2$	(3)
$2+2$	(2)
$1+3$	(3)
4	(1)

WIGNER gave the recursive formula for the degree of degeneracy, given here in brackets for $N = 4$.

In December 1926 HEISENBERG also stated the systems of terms for $N = 3$, but including the spin.[5] For two possibilities α and β

for the individual spin there were four symmetrical combinations of spin:

$$\alpha\alpha\alpha \qquad \alpha\alpha\beta+\alpha\beta\alpha+\beta\alpha\alpha \qquad \alpha\beta\beta+\beta\beta\alpha+\beta\alpha\beta \qquad \beta\beta\beta$$

The number of particles is indicated by the order in which the expressions appear. There are two degenerate and no anti-symmetrical combinations. An orbital state *aaa* which is necessarily symmetrical could not be extended to an antisymmetrical one, which corresponded to the PAULI principle. An orbital state *aab* could be antisymmetrized by two combinations of spin and thus gave a doublet. A state *abc* could be antisymmetrized by means of $\sum(-1)^P a(1)b(2)c(3)$ with the four combinations of spin given above. This would lead to a 'quartet'. Alternatively, using another combination it could be antisymmetrized by two different spin combinations, giving a doublet. In May 1927 Friedrich HUND gave the enumeration of the multiplicities for arbitrary numbers of electrons. He was able to explain the symmetry relationships of the orbital and spin contributions to the eigenfunctions in terms of a rather less abstract process.[9]

Rotation

In May 1927 WIGNER laid the basis of the general application of the theory of group representations to quantum theory and discovered the correspondence between the irreducible representations and the systems of terms. Rotational symmetry about an axis, for example an atom in a magnetic field, led to the notation

$$\exp(i\,M\alpha)$$

for the transformation of the eigenfunctions under a rotation through an angle α, and it also led to a partitioning of the states according as $M = 0, \pm 1, \pm 2,\ldots$. In the case of additional invariance with respect to reflection in any plane through the axis of rotation the same eigenvalue was obtained for $M = \pm\Lambda$. The representations (7) which could now be written as

$$\begin{pmatrix} \cos\Lambda\alpha & -\sin\Lambda\alpha \\ \sin\Lambda\alpha & \cos\Lambda\alpha \end{pmatrix} \qquad \begin{pmatrix} 1 & 0 \\ 0 & -1 \end{pmatrix}$$

held generally, and the states were partitioned according to their behaviour under rotation for $\Lambda = 0, 1, 2,\ldots$, those with $\Lambda = 0$

according to their behaviour under reflection into $+$ and $-$ states, altogether into the types $\Sigma^+, \Sigma^-, \Pi, \Delta,....$ For the case of spherical symmetry, WIGNER found $2L + 1$ eigenfunctions corresponding to each eigenvalue. These behaved like the $2L + 1$ spherical functions of order L, Y_L, and which are multiplied by ± 1 if they are inverted with respect to the centre. Thus we obtain the types:

$$S_+ P_- D_+ F_- ... \qquad S_- P_+ D_- F_+ ... .$$

Dipole transitions are possible only between $+$ and $-$ states (this selection rule was already known empirically). A decrease in symmetry required the reduction of a representation of the higher symmetry into irreducible representations of the lower symmetry. Thus terms with quantum number L gave rise to the terms $\Lambda = 0, 1,..., L$ or $M = -L, -L + 1,..., L$.

The spin did not fit into these representations of the rotation group. Its inclusion was not possible until the development of the two-valued representations of the rotation groups which are given by the numbers $M = \pm\frac{1}{2}, \pm\frac{3}{2},...$ for axial symmetry and by the numbers $J = \frac{1}{2}, \frac{3}{2},...$ for spherical symmetry. These two-valued representations of the rotation group had been discovered by Hermann WEYL in 1925. He used them to describe the atomic states with spin in his 1928 book and had probably already done so in his lecture course in the winter semester of 1927–8.

The definitive version of these applications of group theory to quantum mechanics was given in books by WEYL (1928), WIGNER (1931) and B. L. VAN DER WAERDEN (1932).[10]

The Equivalence of the Wave and Particle Pictures

Stimulated by the wave representations and in the hope of obtaining an intuitive understanding of quantum phenomena, SCHROEDINGER had been led to the discovery of his equation. However, when the equation was interpreted by BORN, PAULI, DIRAC, JORDAN and HEISENBERG, the wave model was rather pushed into the background. These workers saw the SCHROEDINGER equation as a convenient form of expressing the quantum modification of classical particle mechanics. BOHR, however, believed in a kind of equivalence in status of the wave and particle approaches.

Physicists were more familiar with the wave particle duality of light. After all, the achievements of the theory of light waves had been known for well over a century. In quantum theory DIRAC (August 1926) was able to deal with the phenomena of the absorption and emission of light by atoms induced by a radiation field by assuming that quantum mechanics held for atoms and by modelling the light field as an external disturbance by a periodic potential —in other words, by using classical wave theory.[6] As a result of this he developed a perturbation theory for time-dependent phenomena. However, DIRAC was unable to explain the spontaneous emission of light by excited atoms using this approach. He succeeded in doing so in February 1927 by establishing a logical quantum theory of the interaction of light and matter in which the light field was represented by quantized harmonic oscillators.[11] The amplitudes of the light field thus became q numbers or matrices. DIRAC used the complex oscillation quantities constructed out of q and p for light waves (see the Appendix). The equations of motion for these quantities were $\dot{b} = -i\omega b$, $\dot{b}^* = i\omega b^*$. It was possible to express b and b^* in the form

$$\left. \begin{array}{l} b = e^{-i\vartheta}\sqrt{N} = \sqrt{N+1}\,e^{-i\vartheta} \\ b^* = \sqrt{N}\,e^{i\vartheta} = e^{i\vartheta}\sqrt{N+1} \end{array} \right\} \tag{8}$$

using q numbers N and q numbers ϑ. In the matrix representation b was written in the following form, analogous to HEISENBERG's matrix:

$$b = \begin{pmatrix} 0 & \sqrt{1}e^{-i\vartheta_1} & 0 & 0 & \dots \\ 0 & 0 & \sqrt{2}e^{-i\vartheta_2} & 0 & \dots \\ 0 & 0 & 0 & \sqrt{3}e^{-i\vartheta_3}\dots \\ \cdot & \cdot & \cdot & \cdot & \cdot \end{pmatrix}$$

$$= \begin{pmatrix} 0 & e^{-i\vartheta_1} & 0 & 0 & \dots \\ 0 & 0 & e^{-i\vartheta_2} & 0 & \dots \\ 0 & 0 & 0 & e^{-i\vartheta_3}\dots \\ \cdot & \cdot & \cdot & \cdot & \cdot \end{pmatrix} \begin{pmatrix} 0 & 0 & 0 & \dots \\ 0 & \sqrt{1} & 0 & \dots \\ 0 & 0 & \sqrt{2}\dots \\ \cdot & \cdot & \cdot & \cdot \end{pmatrix}$$

$$= \begin{pmatrix} \sqrt{1} & 0 & 0 & \dots \\ 0 & \sqrt{2} & 0 & \dots \\ 0 & 0 & \sqrt{3}\dots \\ \cdot & \cdot & \cdot & \cdot & \cdot & \cdot \end{pmatrix} \begin{pmatrix} 0 & e^{-i\vartheta_1} & 0 & 0 & \dots \\ 0 & 0 & e^{-i\vartheta_2} & 0 & \dots \\ 0 & 0 & 0 & e^{-i\vartheta_3}\dots \\ \cdot & \cdot & \cdot & \cdot & \cdot & \cdot & \cdot \end{pmatrix}$$

It followed from (8) that $b*b = N$, with the eigenvalues $0, 1, 2,\dots$, and that $bb* = N + 1$ with the eigenvalues $1, 2, 3,\dots$. For b and $b*$ the commutation relation $bb* - b*b = 1$ held. If one distinguished between the various eigenvalues, this gave:

$$b_r b_s^* - b_s^* b_r = \delta_{rs} \tag{9}$$

though b_r remained interchangeable with b_s and b_r^* with b_s^*; b_r and $i\hbar b_r^*$ could be regarded as canonically conjugate.

DIRAC applied his model with the quantities b and $b*$ to matter waves as well. He showed that it was possible to represent a BOSE gas in this way. JORDAN noticed (July 1927 in Copenhagen) that there was still one degree of freedom remaining in the rules for the wave amplitudes $b = \exp(-i\vartheta)\sqrt{N}$, $b* = \sqrt{N} \exp(i\vartheta)$.[12] If N was required to have the eigenvalues $0, 1, 2,\dots$, it was necessary to put $bb* - b*b = 1$ as DIRAC had done in accordance with $i(pq - qp) = \hbar$. It was, however, also possible to require the sole eigenvalues 0 and 1, which implied that

$$b = e^{-j\vartheta}\sqrt{N} = \sqrt{1-N}\,e^{-i\vartheta}$$
$$b* = \sqrt{N}\,e^{i\vartheta} = e^{i\vartheta}\sqrt{1-N}$$

so that

$$bb* = 1 - N$$
$$b*b = N$$

and

$$bb* + b*b = 1.$$

The quantity b^2 contains a factor $\sqrt{N(1-N)}$ with the eigenvalue 0. Thus $\sqrt{N}$ and $\sqrt{1-N}$ could be replaced by N and $1-N$. Distinguishing between the various wave types meant putting

$$\left.\begin{matrix} b_r b_s^* + b_s^* b_r = \delta_{rs} \\ b_r b_s + b_s b_r = 0 \\ b_r^* b_s^* + b_s^* b_r^* = 0 \end{matrix}\right\} \tag{10}$$

The two possibilities for the eigenvalues of the quantities N_r corresponded to the number of particles in BOSE and FERMI statistics. JORDAN was thus entitled to express his hope of a quantum wave theory of matter in which the numbers of particles would be the occupation numbers N_r of the discrete quantum wave states.

This hope was fulfilled in a paper by JORDAN and O. KLEIN (October 1927 in Copenhagen).[13] It was possible to eliminate the component of U that depended on the interaction by means of the substitution

$$U(x) = U_e(x) - e \int \frac{\varphi^*(x')\varphi(x)}{|x - x'|}\, d\tau$$

in the equations for a field theory of matter

$$\left(-\frac{\hbar^2}{2m}\Delta - eU - i\hbar\frac{\partial}{\partial t}\right)\varphi = 0$$

$$\left(-\frac{\hbar^2}{2m}\Delta - eU + i\hbar\frac{\partial}{\partial t}\right)\varphi^* = 0$$

$$\Delta U - 4\pi e\varphi^*\varphi = 0.$$

By means of the expansion $\varphi = \sum b_r u_r(x)$ in terms of eigenwaves in the external potential U_e it was possible, up to a point, to treat the field as a system with the Hamiltonian:

$$H = \sum b_r^* b_r E_r + \frac{e^2}{2}\sum A_{rskl} b_r^* b_s^* b_k b_l.$$

The first term corresponds to uncoupled eigenwaves and the second to the coupling through $U - U_e$. Now the b_r were q numbers. With the aid of the commutation relations (9) it was possible to transform the equations for each fixed value $N = \sum N_r$ into equations for the N_r. It was, moreover, possible to transform the SCHROEDINGER equation for N particles

$$\left\{\sum_{n=1}^{N}\left[-\frac{\hbar^2}{2m}\Delta_n - eU_a(x_n)\right] + \sum_{i<k}\frac{e^2}{r_{ik}} - E\right\}\psi(x_1, x_2\cdots) = 0$$

into these same equations by putting

$$\psi = \sum_{r_1 r_2\cdots} b_{r_1 r_2}\cdots u_{r_1}(x_1)u_{r_2}(x_2)\cdots$$

where the coefficients b depended symmetrically on the r_i. This provided the equivalence of the quantum eigenwaves and the symmetrical solutions of the quantized N-particle system. In the same paper JORDAN and KLEIN showed that the commutation relations for the field quantities could also be written in the form

$$\varphi(x)\varphi^*(x') - \varphi^*(x')\varphi(x) = \delta(x - x')$$

$$\varphi(x)\varphi(x') - \varphi(x')\varphi(x) = 0$$

$$\varphi^*(x)\varphi^*(x') - \varphi^*(x')\varphi^*(x) = 0.$$

JORDAN and WIGNER carried out the corresponding calculations for eigenwaves with the commutation relations (10) in January 1928 and were able to demonstrate their equivalence with the antisymmetrical solutions of the N-particle system.[14] This established the symmetry between the wave and particle pictures of matter. The quantization of the matter field and that of material particles led to the same quantum mechanics, and thus finalized non-relativistic quantum theory. The field model was, however, to be of still greater significance in relativistic quantum theory.

The Incorporation of Spin

The spin of the electron had hitherto been taken into account only in so far as it had no consequence beyond the PAULI principle. The magnetic forces that it exerted were neglected. This gap was closed in 1927 in a non-relativistic approximation.[15] A theory that included the spin had to have three properties: (i) every electron had two possible directions in a magnetic field; (ii) the spin had an angular momentum $\hbar/2\pi$; (iii) the spin had a magnetic moment $\hbar e/2mc$ (the ratio of the magnetic moment to the angular momentum was twice as large as in the case of an orbiting charge).

C. G. DARWIN explained the existence of two possible directions in February and July 1927 by means of a SCHROEDINGER equation valid for a two-component quantity ψ and in which the magnetic energy in a field of given direction was taken as different for the two components of ψ corresponding to the direction of the field. The angular momentum properties were explained by PAULI in May 1927 by giving the operators that corresponded to the components of angular momentum P_x, P_y and P_z. These acted on the two components of ψ and satisfied the commutation relations for angular momentum components. These were the 'PAULI spin matrices' σ_x, σ_y, σ_z:

$$\begin{pmatrix} 0 & 1 \\ 1 & 0 \end{pmatrix} \quad \begin{pmatrix} 0 & -i \\ i & 0 \end{pmatrix} \quad \begin{pmatrix} 1 & 0 \\ 0 & -1 \end{pmatrix}$$

which when multiplied by $\hbar/2$ corresponded to the components of angular momentum. The operator for the x-component of the angular momentum of an electron was thus

$$\frac{\hbar}{i}\left(y\frac{\partial}{\partial z} - z\frac{\partial}{\partial y} \right) + \frac{\hbar}{2}\sigma_x.$$

The magnetic energy that resulted from the spin could be expressed by the operator

$$\frac{e\hbar}{2mc}\boldsymbol{B}\boldsymbol{\sigma}.$$

For N electrons ψ was a quantity with 2^N components. VON NEUMANN and WIGNER applied PAULI's theory of spin to atomic spectra. They also drew on WEYL's two-valued representations of the rotation group. The irreducible representations corresponded to the J-values (December 1927). The two men derived the structure of the multiplicity system, the selection and intensity rules (March 1928) and the g-formula of the anomalous ZEEMAN effect (June 1928). In the meantime DIRAC had announced a relativistically invariant theory of spin (see Chapter 15).

[1] E. FERMI, Z. Phys. **36**, 902 (1926)
[2] W. PAULI, Z. Phys. **41**, 81 (1927)
[3] L. H. THOMAS, Proc. Cambr. Philos. Soc. **23**, 542 (1927)
[4] E. FERMI, Z. Phys. **48**, 73, **49**, 550 (1928), Leipziger Vorträge 1928, p. 95
[5] W. HEISENBERG, Z. Phys. **38**, 411, **39**, 499 (1926), **41**, 239 (1927)
[6] P. A. M. DIRAC, Proc. Roy. Soc. **112**, 661 (1926)
[7] D. M. DENNISON, Proc. Roy. Soc. **115**, 483 (1927)
  F. HUND, Z. Phys. **42**, 93 (1927)
[8] E. WIGNER, Z. Phys. **40**, 492, 883, **43**, 642, **45**, 601 (1927)
[9] F. HUND, Z. Phys. **43**, 788 (1927)
[10] H. WEYL, Gruppentheorie und Quantenmechanik, Leipzig 1931
  E. WIGNER, Gruppentheorie und ihre Anwendung auf die Quantenmechanik der Atomspektren, Braunschweig 1931
  B. L. VAN DER WAERDEN, Die gruppentheoretische Methode in der Quantenmechanik, Berlin 1932
[11] P. A. M. DIRAC, Proc. Roy. Soc. **114**, 243 (1927)
[12] P. JORDAN, Z. Phys. **44**, 473 (1927)

[13] P. JORDAN and O. KLEIN, Z. Phys. **45**, 251 (1927)

[14] P. JORDAN and E. WIGNER, Z. Phys. **47**, 631 (1928)

[15] C. G. DARWIN, Nature **119**, 282, Proc. Roy. Soc. **116**, 227 (1927)
W. PAULI, Z. Phys. **43**, 601 (1927)
J. V. NEUMANN and E. WIGNER, Z. Phys. **47**, 203, **49**, 73, **51**, 884 (1928)

14. APPLICATIONS OF QUANTUM MECHANICS

Prediction Methods

WITH the development of the probabilistic interpretation and of transformation theory the principles of quantum mechanics were largely known by the end of 1926. Since the spring of that year the SCHROEDINGER equation had provided a neat method of solving simple problems. This situation brought about a flood of applications and stimulated the development of practical methods of calculation in about 1927.

For mechanical systems that were not quite so easy to handle, SCHROEDINGER's perturbation method of May 1926 was at first adopted. This assumed a Hamiltonian of the form

$$H = H^{(0)} + \lambda H^{(1)}$$

and a solution of the form

$$\psi = \psi^{(0)} + \lambda \psi^{(1)} + \lambda^2 \psi^{(2)} + \cdots$$

The function $\psi^{(0)}$ is a solution of the 'unperturbed' problem $H^{(0)}$. The method amounted to an expansion of the desired function ψ in terms of the complete orthogonal system of the eigenfunctions of $H^{(0)}$. Attention was frequently confined to the solutions of $H^{(0)}$ which belonged to the same eigenvalue $E^{(0)}$. $\psi^{(1)}$ was then expanded in terms of the incomplete system of these functions and this gave a prediction of the splitting of the energy $E^{(0)}$ as a result of the perturbation $H^{(1)}$. The representation in terms of the eigenfunctions of $H^{(0)}$ could already be deduced to a large extent from symmetry arguments. Thus, for example, John C. SLATER, who had already (in 1926) made a similar calculation in the framework of the old quantum theory was able (in 1929) to revise his argument. This time he started from a single electron approximation which described the state of an atom in terms of the quantum numbers $n_1, l_1, n_2, l_2, \ldots$ of the individual electrons.[1] Treating the interaction of the electrons as a perturbation he was able to express the energies of a multiplet corresponding to the quantum numbers n_i, l_i of single electrons using only a few parameters. He thus derived relationships between the distances of the multiplets which belonged to $p^2(^3P^1D^1S)$, p^3, p^4, d^2, d^3,..., d^8 (equivalent electrons) and to p^5p.... He avoided the difficulties of group theory by counting the spin as one of the variables of the single electron functions and writing the required antisymmetric approximation as the determinant of the single electron functions:

$$\psi^{(0)} = \frac{1}{\sqrt{N!}} \begin{vmatrix} a(1) & a(2) & a(3) & .. \\ b(1) & b(2) & b(3) & .. \\ c(1) & c(2) & c(3) & .. \\ . & . & . & .. \end{vmatrix}$$

He then combined the various orbital functions and spins corresponding to n_i, l_i, applying spherical symmetry. The old system of enumeration (Chapter 9) was thus given a quantitative extension.

In practice these already amounted to expansions in terms of incomplete systems of function. The HEITLER-LONDON prediction for the hydrogen molecule[2] led to a second type of model, in terms of a small number of prescribed functions. It amounted to an approximation of the desired function in the form

$$\psi = \sum c_n u_n \tag{1}$$

in terms of given functions u_n, and thus to the problem of determining the 'best' combination c_n. To do this one could put (1) into the SCHROEDINGER equation and obtain equations for the c_n. It amounted to the same thing if the c_n were chosen that gave a particular integral an extreme value for all functions of the form (1). This was the same integral that would have led to the exact solution of the SCHROEDINGER equation by allowing arbitrary functions ψ.

This brings us to the very successful variational method. The desired function is taken to be a simple analytic function

$$\psi = u(x, c_1, c_2 ...)$$

with a number of as yet unknown parameters c_1, c_2,..., and these parameters are determined mostly by trial and error in such a way that the integral corresponding to the SCHROEDINGER equation attains a minimum. G. W. KELLNER used this method to calculate the helium ground state in June 1927, by putting

$$\psi = f(r_1)f(r_2)[1 + \alpha g(\vartheta) + \beta h(\vartheta) + \gamma k(r_1)k(r_2)]$$

r_1 and r_2 denoting the distances of the electrons from the nucleus, ϑ being the angle between these distances and f, g, h and k simple given functions. α, β and γ were determined by variational methods. E. A. HYLLERAAS was one physicist who became particularly skilled in this method. In February 1929 he obtained a very precise energy value for the helium ground state using

$$\psi \sim e^{-\alpha(r_1 + r_2)}[1 + \beta r_{12} + \gamma r_{12}{}^2 + \delta(r_1 - r_2)^2 + \varepsilon(r_1 + r_2) + \zeta(r_1 + r_2)^2].$$

He went on to calculate a vast number of atomic energy states.[3]

The method used by D. R. HARTREE (October 1927) turned out to be very suitable for predictions of atoms with higher atomic numbers.[4] A first approximation to the value of the potential was used to solve the SCHROEDINGER equation for single electrons. An improved value of the potential was calculated from the solutions, as was the electron density. These in turn were used to solve SCHROEDINGER equations for the single electrons and so on, until the repetition of the process caused no noticeable change. This led to the determination of a 'self-consistent' field. J. A. GAUNT and J. C. SLATER showed in 1928 that the solution of the SCHROE-DINGER equation using the substitution

$$\psi = a(1)b(2)c(3)\ldots \tag{2}$$

where 1, 2, 3,... denote the co-ordinates of the electrons and a, b, c,... were functions to be determined, led to precisely the HARTREE equations. In 1930 V. FOCK improved the method by allowing sums of products instead of (2) and thus took account of the more or less antisymmetric relationship between the functions of the single electrons, corresponding to the multiplet terms.

When we come to the application of the SCHROEDINGER equation to the collision of particles with atoms we see two lines of argument developing. In the first the SCHROEDINGER equation was solved rigorously subject to a restriction to elastic collisions. By spherical symmetry of the force field and the axial symmetry of the phenomenon it was possible to calculate separately the parts of the SCHROEDINGER function which corresponded to the axisymmetrical spherical harmonic functions $P_l(\cos \delta)$. A phase difference between the emitted and received waves was obtained for each part. Thus it was possible to explain for example the RAMSAUER effect (Chapter 11).[5] The second line of argument was to regard the influence of the atom as relatively small. These were extensions of BORN's work of 1926 (Chapter 12). DIRAC developed a perturbation method for time-dependent perturbations in the context of his first application of the SCHROEDINGER equation in August 1926. Using the series $\psi = \sum a_n u_n$ he set up equations for the derivatives $\dot{a}_n$. We have already learned of their application to absorption and to the component of the emission that is due to the incoming radiation.

Spreading Eigenfunctions

A characteristic difference between classical mechanics and quantum mechanics is the way in which the eigenfunction in quantum mechanics can spread beyond the limits $V = E$ of the classical motion. For a barrier $V > E$ between two regions where $V < E$ (Figure 22) there are two types of motion in classical

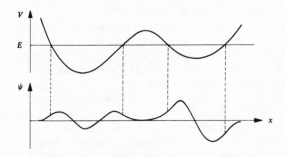

FIGURE 22: THE POTENTIAL BARRIER

mechanics, each belonging to one of the potential troughs. The wave function that belongs to E, however, spreads from the one trough into the other, in the case of higher or wider barriers by only a very slight amount. *A barrier is not completely impenetrable.* In fact it allows what was later to be called the 'tunnel effect'. HUND recognized the importance of this for the theory of molecules in November 1926. In August 1927 J. Robert OPPEN-HEIMER recognized its importance in the theory of an atom in an electric field. They were followed by NORDHEIM in December 1927, who dealt with the case of electron emission by metals, and in July and August 1928 for α-decay of heavy atomic nuclei by R. W. GURNEY and E. U. CONDON and also George GAMOW.[6]

HUND was able to interpolate the states of the electrons for a biatomic molecule as states in a system of two troughs lying between the limiting cases of separated and united atoms. For two identical atoms the states of the separated atoms split into one that was symmetrical about the mid-plane and one that was anti-symmetrical about it. R. S. MULLIKEN's experimental investigations of simple molecules were particularly important for the classification of the molecular spectra. In 1928 he showed that the single electron approximation was quite useful. BORN and OPPENHEIMER's series expansion in terms of the electron/nucleus mass ratio in 1927 showed the interaction of rotation, oscillation and electron motion in a molecule. R. KRONIG's contributions of December 1927 and June 1928 dealt with the rotation terms precisely. In May 1927 HUND demonstrated the consequences for atomic nuclei of the spreading of the wave functions: in particular that beat frequencies could arise which lay between atomic and cosmic orders of magnitude.[7] OPPENHEIMER, in August 1927, gave a general mathematical treatment of the spreading between two potential troughs and, as noted, applied the results to a hydrogen atom in an external electric field. NORDHEIM showed (December 1927) the influence of a surface layer idealized in terms of a barrier for the thermal emission of electrons by metals. NORDHEIM and FOWLER (March 1928) treated the problem of spreading for electron emission in a strong electric field. The most impressive work of this kind was surely the explanation of α-decay for radioactive nuclei as the penetration of a potential barrier between a force field which tended to hold the α-particles in the interior of the nucleus, and the external region. This was done simultaneously by GURNEY and CONDON (July 1928) and by GAMOW (August 1928).

Chemistry

One of the greatest successes of quantum theory was the absorption of theoretical chemistry into the thinking of physicists. The treatment of the chemical bond in terms of theoretical physics naturally presupposes a theory of molecules. It would thus have been impossible without the new quantum mechanics and without the SCHROEDINGER equation. The single electron approximation which at first proved very useful for the description of molecular states was, however, not so well suited to the treatment of the chemical bond. In this case it was in fact a question of the relationship of the molecular states to the states of separate atoms, and, in this limit, the (neglected) finer interaction between the electrons became greater than the bonds between the atoms.

A basic understanding for the chemical bond was achieved by HEITLER and LONDON in June 1927 when they estimated the energy of the ground state of a hydrogen molecule by approximating the eigenfunction

$$u(1,2) \sim a(1)b(2) \mp b(1)a(2) \tag{3}$$

in terms of the ground states a and b of the two atoms (without spin; 1 and 2 denote the co-ordinates of the two electrons).[2] The expression they obtained for the approximate value of the energy was

$$E = 2E_0 + \frac{C \mp A}{1 \mp S} \tag{4}$$

where E_0 was the energy of a hydrogen atom, S was an integral which gave the overlapping of the functions a and b, and C was a magnitude $(C < 0)$ which could be interpreted as a Coulomb interaction between the two 'electron clouds' and the nuclei. The 'exchange integral' A was a quantity which could only be explained in terms of quantum mechanics. The 'chemical bond' between the two atoms corresponded to the symmetrical relationship between the atomic eigenfunctions a and b in (3) which (for $A < 0$) led to a considerable reduction in the energy, C being quite small for large distances. The antisymmetric relationship in (3) led to a minus sign in front of A and thus to a higher energy. Instead of a symmetrical relationship between the electron orbit functions it was possible to refer to the antisymmetric relationship between their spins. The corresponding prediction for He_2 gave no bond at all, as the path functions are already related symmetrically for the separate atoms and the spins antisymmetrically. The HEITLER-LONDON approximation to the molecular state using a very incomplete system of functions $a(1)b(2)$ and $b(1)a(2)$, which were not even orthogonal, contributed to the recognition of this type of rough approximation with a small number of prescribed functions.

In December 1927 LONDON saw how the behaviour of a quantum

system could be interpreted as a model of chemical valency.[8] If the electrons in an atom or in a group of atoms are already symmetrically related (forming a singlet term) the inclusion of a second atom or a second group of atoms does not lead to any new symmetric combination nor to any splitting of the energy ($^1T + {}^1T \rightarrow {}^1T$, $^2T + {}^1T \rightarrow {}^2T$). No valency is present. If the electrons in an atom or in a group of atoms are not symmetrically related, giving ^2T, ^3T, the approach of a partner with likewise non-symmetrical electrons causes splitting, as for example

$$^2T + {}^2T \rightarrow {}^1T, {}^3T \qquad {}^3T + {}^3T \rightarrow {}^1T, {}^3T, {}^5T \qquad {}^2T + {}^3T \rightarrow {}^2T, {}^4T.$$

The reduction in the multiplicity by 1 denoted saturation of a valency, a reduction by 2 the saturation of two valencies. The triple bond in N_2 was seen as a transformation $^4S + {}^4S \rightarrow {}^1\Sigma$. LONDON's 'theory of spin valence' gave a correct indication of one fundamental point. It holds, however, only so long as it is possible to neglect the spatial degeneration of the electron paths within the atom. It is thus merely a model of valence behaviour.

HERZBERG, who was convinced of the value of the single electron approximation (in terms of eigenfunctions of individual electrons in the molecular force field) by his research into molecular spectra, distinguished between 'slack' and 'binding' states in biatomic molecules according to whether or not the bringing together of the atoms led to a new nodal surface of the wave function.[9] The saturation of a valency consisted in the fact that an electron of one atom went into a binding state with an electron of a neighbouring atom.

The removal of the degeneracy of p states (which depends on the spherical symmetry of the atoms) makes a fundamental contribution to the theory of the chemical bond. Moreover, if the bond is sufficiently strong the distinction between s and p states loses all meaning and, for example, the four external electrons of a carbon atom all enter into equivalent bonds with neighbouring atoms. This was recognized by Linus PAULING in March 1928. The precise argument which was announced in his paper was, however, not completed until the beginning of 1931, shortly after work done by J. C. SLATER.[10] Both writers regarded it as fundamental for chemical bonds that the wave function of an electron should be directed towards its binding partner. Two p electrons (p^2) can then bind two partners most strongly if the directions they make with it form a right angle. Three p electrons (p^3) can bind three partners most strongly if the three directions form a rectangular trihedron. If the bond is sufficiently strong for the distance between s and p terms of an atom to be treated as small it is possible to form directed atomic functions with suitable combinations of s and p functions which are suitable to the description of the bond. Three functions of this kind form a maximal bond if the lines joining them with the partner lie in a plane and make angles of 120° with each other. Four functions bind in this way if the directions correspond to an arrangement in the form of a regular tetrahedron. In August 1931 SLATER provided the mathematical basis for his theory of directed valence by approximating the eigenfunction of the molecule in terms of the sum of a number of products of eigenfunction of single electrons. The phenomena of valence angles could also be explained in terms of the single electron approximation of the molecular states. The fact, known to chemists, that the components of a molecule that are connected by a simple bond can be rotated about this bond with respect to each other and that a double or triple bond is not rotatable was explained by Erich HUECKEL at the beginning of 1930. He attributed it to the involvement of π electrons with a wave function that was not axisymmetric, in the multiple bond. The particular conditions for benzene and other 'aromatic' compounds were connected in 1931 with the symmetrical behaviour of π electrons which enter 'non-localized' bond.[11]

Thus by 1931 there were three quantum models of the chemical bond. The LONDON model of spin valence approximated the state of molecule by the many-electron ground states of the separate atoms and thus at first regarded the chemical bond as weak compared with the interaction between the electrons which led to the different multiplets. LONDON saw valences of the partners in a bond as being expressed in the multiplicity of their ground states and the saturation of valences as a consequence of a reduction in the multiplicity as a result of the formation of the molecule. SLATER's generalization of the HEITLER-LONDON approach approximated the state of the molecule in terms of the states of the individual electrons in the separate atoms. It thus introduced the interaction between the electrons and the chemical bond at the same step in the approximation. Those connections of the single electron function were of particular significance in this work which corresponded to the valence lines used by chemists. It was often a good idea to use as single electron functions of the atom those combinations that represented directed valences. The third model, the approximation of the state of the molecule in terms of single electron functions of the molecule, led to their partition into binding and non-binding functions and to a partition of the bonds into localized σ or π bonds and into non-localized ones. In this version the interaction between the electrons was regarded as small as compared with the chemical bond. That was actually often the case in molecules, but it does not hold for the transition to separate atoms.

It was clear that the three versions corresponded to coarse approximations which involved the neglect of a quantity which was not in fact small but which was regarded as less typical. The various interpretations of chemical bonds were later to develop into methods of 'quantum chemistry': the valence bond method (VB), which started out as the VB-LCAO method from the SLATER approximation in terms of linear combinations of atomic functions (linear combinations of atomic orbitals), and the MO method that started from the single electron approximation in the molecule (molecular orbitals) which, as the special MO-LCAO method, approximated the single electron function in the molecule by means of those in atoms.

Electrons in Metals

Already in December 1926 PAULI had recognized FERMI statistics as the essential difference between the electrons in a metal and in a conventional gas. Its corollary of a temperature-independent paramagnetic susceptibility marked the beginnings of a quantum treatment of the properties of metals. SOMMERFELD embarked on a comprehensive investigation of the model of the free electrons in a metal (October and December 1927). He calculated the specific heats and the connections between the electrical and thermal conductivities. The cause of the temperature dependence of electrical resistance was seen by FRENKEL (February 1928 in Leningrad) as the fact that the thermal oscillations of the crystal lattice led to a disturbance of the waves that corresponded to the electrons and to a restricted free path length. He made the proportionality of resistance to temperature at least plausible. Almost simultaneously W. V. HOUSTON (March 1928, in Munich at the time) made a more precise prediction of this scattering. This gave a proportionality between the resistance and the temperature for higher temperatures, but it also gave an incorrect result for lower temperatures.[12]

A further difference between the electron gas in a metal and an ordinary gas comes from the distinction between the behaviour of the SCHROEDINGER functions and the motion of classical particles. Lev LANDAU (July 1930 in Cambridge), showed by solving the SCHROEDINGER equation in the presence of a magnetic field that a gas composed of free electrons displayed a weak diamagnetism, which contradicted classical theory.[13]

Hans BETHE (Summer 1928, in Munich) and Felix BLOCH (Summer 1928, in Leipzig) tackled the behaviour of electrons in the crystal lattice of a metal.[14] BETHE explained the interference of electrons on reflection by crystal surfaces. According to BLOCH, translation invariance

$$V(x) = V(x+a)$$

led to a translation quantum number k and so to a representation of the wave functions of the individual electrons in the form

$$\psi = e^{ikx} V_k(x) \qquad V_k(x) = V_k(x+a).$$

BLOCH was able to predict the perturbation of the electron states by a vibrating lattice. The problem led, by a BOLTZMANN-type argument, to an integral equation, the solution of which gave a resistance proportional to the temperature for high values of the temperature. Towards the end of 1929 BLOCH also discovered a solution for low temperatures and a resistance proportional to T^5 caused by the thermal waves. In his 1928 paper BLOCH also treated the approximation of the energy band function $E(k)$ for the case of a strong binding of the electrons to the atomic cores. For a simple cubic lattice

$$E = C + 2K(\cos k_x a + \cos k_y a + \cos k_z a)$$

he was able to demonstrate the differences between the behaviour of a lattice electron and a free electron.

The work of PEIERLS on diamagnetism of the electrons in a metal at low temperatures (1932–3) had considerable consequences. His theorem, that, if $E(k)$ was the energy band function of the metal, the Hamiltonian

$$H = E\left(k + \frac{e}{\hbar} A\right)$$

gave the approximate behaviour of the lattice electrons in an external magnetic field, formed the basis of the subsequent experimental determination of the 'FERMI surfaces' $E(k) = \zeta$ for metal lattices. In all these applications of quantum theory to electrons in metals we have neglected the interaction between the conducting electrons. This interaction was already recognized at the time as the cause of ferromagnetism. FRENKEL (March 1928) deduced from what had been observed for atomic spectra that the parallel alignment of the spins denoted a diminution of the energy. If this compensates for the energy needed to reach the higher electron state now required by the PAULI principle, a ferromagnetism will result.

In May 1928 HEISENBERG tackled the problem by means of an approach to the interaction that was analogous to the HEITLER-LONDON calculation for H_2, and was thus able to explain why ferromagnetism may arise in certain circumstances. BLOCH in turn started from the single electron approximation in June 1929 and gave a quantitative evaluation using FRENKEL's approach.[16]

[1] J. C. SLATER, Phys. Rev. **28**, 291 (1926), Phys. Rev. **34**, 1293 (1929)

[2] W. HEITLER and F. LONDON, Z. Phys. **44**, 455 (1927)

[3] G. W. KELLNER, Z. Phys. **44**, 91 (1927)
E. A. HYLLERAAS, Z. Phys. **48**, 469 (1928), **54**, 347 (1929), **60**, 624, **63**, 291, **65**, 209 (1930)

[4] D. R. HARTREE, Proc. Cambr. Philos. Soc. **24**, 89, 111 (1928)
J. A. GAUNT, Proc. Cambr. Philos. Soc. **24**, 328 (1928)
J. C. SLATER, Phys. Rev. **35**, 210 (1929)
V. FOCK, Z. Phys. **61**, 126 (1930)

[5] H. FAXÉN and J. HOLTSMARK, Z. Phys. **45**, 307 (1927)
L. MENSING, Z. Phys. **45**, 603 (1927)
J. HOLTSMARK, Z. Phys. **48**, 231 (1928), **55**, 437 (1929), **66**, 49 (1930)

[6] F. HUND, Z. Phys. **40**, 742 (1927)
J. R. OPPENHEIMER, Phys. Rev. **31**, 66 (1928)
L. NORDHEIM, Z. Phys. **46**, 833 (1928)
R. W. GURNEY and E. U. CONDON, Nature **122**, 439 (1928), Phys. Rev. **33**, 27 (1929)
G. GAMOW, Z. Phys. **51**, 204 (1928)

[7] M. BORN and J. R. OPPENHEIMER, Ann. Phys. **84**, 457 (1927)
R. KRONIG, Z. Phys. **46**, 814, **50**, 347 (1928)
F. HUND, Z. Phys. **43**, 805 (1927)

[8] F. LONDON, Z. Phys. **46**, 455, **50**, 24 (1928)

[9] G. HERZBERG, Z. Phys. **57**, 601 (1929)

[10] L. PAULING, Proc. Nat. Akad. **14**, 359 (1928), Jn. Amer. Chem. Soc. **53**, 1367 (1931)
J. C. SLATER, Phys. Rev. **37**, 481, **38**, 1109 (1931)

[11] E. HUECKEL, Z. Phys. **60**, 423 (1930), **70**, 204 (1931)

[12] W. PAULI, Z. Phys. **41**, 81 (1927)
A. SOMMERFELD, Naturwiss. **15**, 825 (1927), Z. Phys. **47**, 1, 43 (1928)
J. FRENKEL, Z. Phys. **47**, 819 (1928)
W. V. HOUSTON, Z. Phys. **48**, 449 (1928)

[13] L. LANDAU, Z. Phys. **64**, 629 (1930)
[14] F. BLOCH, Z. Phys. **52**, 555 (1928)
[15] R. PEIERLS, Z. Phys. **80**, 763, **81**, 186 (1933)
[16] J. FRENKEL, Z. Phys. **49**, 31 (1928)
   W. HEISENBERG, Z. Phys. **49**, 619 (1928)
   F. BLOCH, Z. Phys. **57**, 545 (1929)

15. FURTHER DEVELOPMENTS IN QUANTUM MECHANICS

Relativistic Quantum Theory

AS far as principles were concerned, quantum mechanics was complete by 1927. A system was characterized by its Hamiltonian $H(p_i, q_j)$. Physical observables were represented by Hermitian operators in Hilbert space, and the operators corresponding to the canonical variables p and q satisfied the commutation relations. Probabilistic assertions were made as to the numerical values of physical quantities. But all this was non-relativistic quantum mechanics. There was indeed even in classical physics no relativistic mechanics for systems of many particles because of the finite propagation of all interactions. But there was already a relativistic mechanics of an electrically charged particle (charge e) in an electromagnetic field given by the potential U and A, with the equation

$$-\left(\frac{E}{c} - eU\right)^2 + (p - eA)^2 + m^2 c^2 = 0 \tag{1}$$

connecting kinetic energy and momentum. It was thus reasonable to anticipate the development of a relativistic mechanics for a single particle or for uncoupled particles. As it was possible to construct non-relativistic quantum mechanics on the basis of the intuitive field model, and a field model worked for the relativistic domain of classical physics, it was also reasonable to anticipate a relativistic quantum field theory. Quantum electrodynamics had in fact already begun, with the treatment of radiation in terms of quantized oscillators, which also gave an indication of how a quantum field theory of matter could be developed.

The relativistic quantum theory of a particle, in the form of the 'DIRAC theory' of the electron, led quite astonishingly to a theory of spin and to the concept of an 'anti-electron'. Generalized quantum field theories were soon designed. But serious difficulties emerged in their detailed treatment. An important feature of relativistic quantum field theory was that the number of material particles did not need to be constant. It was possible for particles to be created or destroyed or to change into other particles. This quantum field theory, which was an extension of quantum mechanics, was thus the bequest made in about 1925 by the era of theoretical physics to the succeeding era—that of high-energy physics, or elementary particle theory. This new type of physics was characterized by the creation and annihilation of particles.

Relativistic Wave Equations

The equation for the four-vector $(E/c, \boldsymbol{p})$ composed of the energy and momentum of a particle:

$$-\frac{E^2}{c^2} + \boldsymbol{p}^2 + m^2 c^2 = 0 \tag{2}$$

and its development (1) and also the DE BROGLIE equation for the four-vector $(\omega/c, \boldsymbol{k})$ with components frequency and wave number suggested a wave equation for matter[1]

$$\left\{ -\left(\frac{i\hbar\partial}{c\partial t} - eU\right)^2 + \left(\frac{\hbar}{i}\nabla - e\boldsymbol{A}\right)^2 + m^2 c^2 \right\}\psi = 0. \tag{3}$$

It was possible to interpret it as an equation for an intuitive field of matter, or as an equation for a particle with rest-mass m and charge e. It is to be found in a paper by KLEIN (April 1926) in a five-dimensional unification of gravitation and the electromagnetic field. SCHROEDINGER also gave it in June 1926, and he pointed out that it led to the old SOMMERFELD formula for the fine structure of the hydrogen lines. It is also given by other authors. Walter GORDON derived the expressions for charge and current from the corresponding variational principle (September 1926):

$$s_\alpha \sim \frac{\hbar}{2i}\left(\psi^* \frac{\partial \psi}{\partial x^\alpha} - \psi \frac{\partial \psi^*}{\partial x^\alpha}\right) - eA_\alpha \psi^* \psi \quad \alpha = 0, 1, 2, 3. \tag{4}$$

Equation (3) was subsequently often to be referred to as the KLEIN-SCHROEDINGER-GORDON equation. In December 1926 KLEIN derived the non-relativistic SCHROEDINGER equation from it by setting

$$\psi = e^{-\frac{i}{\hbar}mc^2 t} \varphi(\boldsymbol{x}, t)$$

where φ changes slowly with respect to time. In December 1926 SCHROEDINGER demonstrated the validity of the conservation law for the sum of the energy-momentum tensors of matter and of the electromagnetic field. He saw it as a problem that the action of matter upon itself is not taken into account in these equations. He did not yet recognize that equation (3) was a single particle equation.

In 1928 WEYL demonstrated the invariance of the electromagnetic potentials with respect to a particular type of transformation (called ' gauge invariance '), in other words, with respect to the simultaneous substitution of

$$\psi \rightarrow \psi e^{i\frac{e}{\hbar}\Phi}$$

and of

$$A_\alpha \rightarrow A_\alpha + \frac{\partial \Phi}{\partial x^\alpha}$$

which is ensured by the combination

$$\frac{\hbar}{i}\frac{\partial}{\partial x^\alpha} - eA_\alpha \qquad (\alpha = 0, 1, 2, 3).$$

At the time it was not realized that equation (4) with the two signs for the density s_0 of electric charge also contained the possibility of the simultaneous creation of positively and negatively charged matter. Equation (3) was on the whole rather forgotten once DIRAC had developed a different LORENTZ-invariant theory.

Dirac's Equation for the Electron

Equation (3) gave no prediction of the spin. Neither did it fit into the general pattern of quantum mechanics and its physical interpretation, which appeared to require an equation of the type

$$H\psi - i\hbar\dot{\psi} = 0.$$

DIRAC succeeded in January 1928 in deriving a LORENTZ-invariant equation of this type which surprisingly also comprised the spin. In order to obtain an equation which contained only the first derivative with respect to time, he factorized (2) as

$$\left(-\frac{E}{c} + \boldsymbol{\alpha}\boldsymbol{p} + \alpha_4 mc\right)\left(\frac{E}{c} + \boldsymbol{\alpha}\boldsymbol{p} + \alpha_4 mc\right) = 0, \qquad (5)$$

where it must necessarily now hold that

$$\alpha_\mu \alpha_\nu + \alpha_\nu \alpha_\mu = 2\delta_{\mu\nu} \qquad (\mu, \nu = 1,2,3,4).$$

He was able to satisfy the equations by using the four 4×4 matrices

$$\begin{pmatrix} 0 & 0 & 0 & 1 \\ 0 & 0 & 1 & 0 \\ 0 & 1 & 0 & 0 \\ 1 & 0 & 0 & 0 \end{pmatrix} \begin{pmatrix} 0 & 0 & 0 & -i \\ 0 & 0 & i & 0 \\ 0 & -i & 0 & 0 \\ i & 0 & 0 & 0 \end{pmatrix} \begin{pmatrix} 0 & 0 & 1 & 0 \\ 0 & 0 & 0 & -1 \\ 1 & 0 & 0 & 0 \\ 0 & -1 & 0 & 0 \end{pmatrix} \begin{pmatrix} 1 & 0 & 0 & 0 \\ 0 & 1 & 0 & 0 \\ 0 & 0 & -1 & 0 \\ 0 & 0 & 0 & -1 \end{pmatrix}$$

For the case of one electron he therefore set up the equation

$$\left\{ \frac{i\hbar \partial}{c \partial t} + eU + \alpha \left(\frac{\hbar}{i} \nabla + eA \right) + \alpha_4 mc \right\} \psi = 0 \tag{6}$$

taking into account the addition of an electromagnetic potential; the electron charge is now written as $-e$, and ψ now denotes a column of four functions. This was joined by a corresponding equation for the complex conjugate functions. DIRAC proved the LORENTZ-invariance of his equation and demonstrated the conservation of electric charge with the densities

$$\rho \sim -\psi^* \psi \qquad s \sim -\psi^* \alpha \psi$$

ψ^* being written as a row of four functions. By multiplying equation (6) without U or A by the operator

$$-\frac{i\hbar \partial}{c \partial t} + \alpha \frac{\hbar}{i} \nabla + \alpha_4 mc$$

on the left, by (5) this led to the wave equation

$$\frac{\hbar^2}{c^2} \ddot{\psi} - \hbar^2 \Delta \psi + m^2 c^2 \psi = 0.$$

But multiplication on the left of the complete equation (6) by the operator

$$-\left(\frac{i\hbar \partial}{c \partial t} + eU \right) + \alpha \left(\frac{\hbar}{i} \nabla + eA \right) + \alpha_4 mc$$

did not give rise to wave equation (3). In fact, extra terms appeared. They represented an additional energy in the electromagnetic field which described the spin as they corresponded for

The History of Quantum Theory

slow motion to a magnetic moment $he/2mc$. The DIRAC equation thus turned out to be a theory of spin.

Equation (2) allows both positive and negative values of the kinetic energy E. Equation (1) allows positive and negative values of the kinetic energy $E—ceU$; in fact, it allows the regions

$$E - ceU \leq -mc^2.$$
$$E - ceU \geq +mc^2.$$

In classical mechanics it is possible to ignore the negative regions, but in quantum theory, as DIRAC soon noticed, it is also possible for transitions to occur from one region into the other.

There soon followed applications of DIRAC's equation.[3] Thus GORDON and DARWIN immediately derived the SOMMERFELD fine-structure formula, which was thus given a completely new explanation. DARWIN showed the extent to which his and PAULI's non-relativistic spin theory represented an approximation of DIRAC's theory. GORDON showed, a little later, that the electric current in the DIRAC theory could be resolved into a conduction current and a polarization and magnetization current, each of which satisfied conservation laws, and the first of which corresponded to the motion and the second to the spin of the electron. Important among the applications was the prediction of the COMPTON scattering of light by free electrons which was carried out by KLEIN and Y. NISHINA in Copenhagen in August 1928. The predicted deviation from earlier theories appeared to be experimentally verified. Thus, in 1928, the validity of the fine structure formula and of the formula due to KLEIN and NISHINA convinced physicists of the correctness of the DIRAC equation.

It was soon seen that the four-component DIRAC ψ was a particular kind of entity that was neither a scalar nor a vector, nor even a tensor. It was called a spinor. WEYL's 1928 book implicitly contains a kind of spinor analysis. An explicit formulation was given by B. L. VAN DER WAERDEN in June 1929.[4] He wrote the DIRAC equation in a form which we may write for brevity as

$$\text{Der } \psi = \kappa\chi$$
$$\text{Der } \chi = \kappa\psi.$$

where ψ and χ are two-component spinors, and Der ψ is the covariant derivative of ψ.

The Hole Theory

The possibility of a transition to negative values of the kinetic energy was a serious difficulty in the DIRAC theory of the electrons. KLEIN showed in December 1928 that a particularly simple example was the behaviour of a DIRAC wave as it approached a high potential barrier $(-eU > 2mc^2)$.[5] We can understand his argument by means of Figure 23. Distances of mc^2 above and below the value of the potential energy represent the beginnings of possible energy regions for the electron. For an energy that lies on the left in the upper region and on the right in the lower permitted region a wave approaching from the left is partially reflected and partially transmitted to the right. This in fact means that a stream of electrons approaching from the left is partly reflected but partly penetrates as a stream of electrons of negative kinetic energy into the region to the right of the step. The conservation of energy holds as the particles receive a very high potential energy. As a result of exposure to a force field, particles of this kind, with negative kinetic energy and thus with negative mass, undergo an acceleration in the opposite direction to the force. In an electric field they therefore behave like positively charged particles. They do not, however, resemble them in every respect: they are after all negatively charged, because electric charge is conserved.

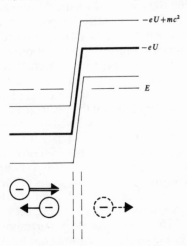

FIGURE 23: THE KLEIN PARADOX

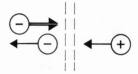

FIGURE 24:   THE DESTRUCTION OF PARTICLES

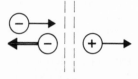

FIGURE 25:   THE CREATION OF PARTICLES

In December 1929 DIRAC showed one way out of the difficulty.[6] As electrons satisfy the PAULI principle, he was able to assume that all states of negative mass were occupied in the usual way. For DIRAC an unoccupied state of negative mass then signified the presence of a particle of positive mass and positive electric charge. He put it equal to the proton, the only known elementary particle of positive charge. It was then necessary to reinterpret the KLEIN paradox. On the right hand of Figure 23 previously unoccupied states of negative mass are filled up, i.e., positively charged particles disappear, as shown in Figure 24. Of course the reverse process can also occur, as shown in Figure 25, the creation of equal numbers of particles of positive and negative charge. The infinite electric charge density which was at first given by the new interpretation had somehow to be eliminated by a new explanation of charge. The asymmetry of the masses of the electron and the proton in the otherwise symmetrical theory seemed to DIRAC to represent a difficulty. He felt that he ought to relate it to a difference in the interaction between electrons or between protons, an interaction which was not yet completely understood. DIRAC explained the scattering of light by an electron in terms of the absorption and emission of a quantum of light and as an intermediate state for which the energy law did not need to hold because it was of such short duration. DIRAC's explanation corresponded to the diagrams, shown in Figure 26, which were not used until very much later. Time increases from left to right; the thick line represents an intermediary proton created simultaneously with an electron and immediately afterwards disappearing with another electron. DIRAC showed that his predictions of the Compton effect contained all these components, and that the component with the intermediary positive particle was essential.

FIGURE 26: CONTRIBUTIONS TO THE COMPTON EFFECT

In February 1930 J. R. OPPENHEIMER showed that the asymmetry of the mass that arose in DIRAC's theory endangered predictions that were in accordance with experimental results. He estimated that the proton and the electron of a hydrogen atom would have to annihilate each other in the space of about 10^{-10} seconds by emitting radiation. He reckoned that electrons and protons were independent particles. The states of negative energy must be occupied for each kind of particle. In March 1930 DIRAC, and in April 1930 TAMM, calculated a high probability for the annihilation of the particles. DIRAC, however, believed that the interaction, which was not yet embraced in his theory, could save the stability of the proton and the electron. However, one year later, in May 1931, he declared the 'holes' to be new unknown particles, 'anti-electrons' which had the same mass as electrons but with charge $+e$. He also regarded the existence of an anti-proton as quite possible.[7]

When he came to complete his contribution to the *Handbuch der Physik* which appeared in 1933,[8] PAULI still regarded DIRAC's way out of the difficulty of the states of negative energy as unsatisfactory. Not until the experimental observation of the positron by C. D. ANDERSON in August 1932 was the DIRAC 'hole' theory generally believed.

Quantum Field Theory

The wave-particle dualism opened two approaches to quantum theory. The approach for matter had led from particles to a rigorous non-relativistic quantum mechanics, but this approach was incapable of providing a relativistic mechanics in the general sense. On the other hand, it was possible to construct a quantum field theory equivalent to non-relativistic quantum mechanics from the wave or field approach. As the relativistic region appeared to raise no problems for a field theory, it was reasonable to hope to derive a relativistic quantum theory on the basis of the field approach. To a certain extent, this hope was fulfilled. The concept of a matter field and its quantization which permitted the existence of particles were explained. The possibility of the creation and annihilation of particles emerged as a general characteristic of relativistic quantum theory. The discovery of new elementary particles and the investigation of their transformations fitted into the general scheme of quantum field theory. This occurred because the characteristic phenomena associated with elementary particles—their transformations based on couplings—found their counterparts to a certain extent in the coupling of various fields. Apart from 'electrical' coupling, or the interaction between the electromagnetic field and the DIRAC field (γee), there were, for example, the interactions ($npe\nu$) between neutrons, protons, electrons, and neutrinos which gave β decay and the coupling ($NN\pi$) between nucleons (protons and neutrons and π mesons which all made a fundamental contribution to the forces between the nucleons. The development of this is of interest to us in so far as it represented the realization of concepts of quantum theory.

The earliest quantum field theory was that of light. It began with the quantization of electromagnetic eigenwaves in radiation by EHRENFEST (1906) and DEBYE (1910) and by EINSTEIN's (1909) recognition of the effects of particles and waves in fluctuations of energy and momentum in the radiation field (Chapter 3). The precise quantum theory of the harmonic oscillator with the matrix elements $x_{x,\,n+1} \sim \sqrt{n\;+\;1}$ created a new situation. DIRAC thereupon treated the radiation amplitudes (1927) as q numbers with elements $b_{n,\,n+1} \sim \sqrt{n\;+\;1}$.[9] The absorption of light became the annihilation of light quanta with a probability proportional to $|b_{n,\,n-1}|^2 \sim n$, the emission of light to the creation of light quanta with a probability that was proportional to $|b_{n,\,n+1}|^2 \sim n\;+\;1$, which contained both induced and spontaneous emission. The terms

$$b_i a_k a_l^* \qquad a_l a_k^* b_i^*$$

appeared in the interaction where a_k were the coefficients in the series for the wave functions of matter. These gradually came to be understood as symbols for the annihilation of a light quantum, the annihilation of an electron in the state k, the creation of an electron in the state l, the annihilation of an electron in the state l and the creation of an electron in the state k and of a light quantum. These were later represented by diagrams like Figure 27.

FIGURE 27: THE ABSORPTION AND EMISSION OF A QUANTUM OF LIGHT

But even this did not amount to a general quantum electro-dynamics. This of course had to be a theory of the field described by the electromagnetic potentials U and A. JORDAN and PAULI in December 1927 gave LORENTZ-invariant commutation relations for the electromagnetic field quantities $\beta_{ik}(i, k = 0, 1, 2, 3)$ without interaction.[10] Field quantities at the space-time points x, x' which did not act upon each other had to commute. The quantity

$$B_{ik}(x)\,B_{lm}(x') - B_{lm}(x')\,B_{ik}(x)$$

could thus be non-zero only if x and x' could be connected by a light ray. While using the Fourier expansion of the field quantities, JORDAN and PAULI discovered that

$$B_{ik}(x)\,B_{lm}(x') - B_{lm}(x')\,B_{ik}(x) \sim \Delta_{iklm}(x' - x)$$

where Δ_{iklm} was the derivative of a singular invariant function $\Delta(x - x')$, which was non-zero only if $(x_i - x'_i)(x^i - x'^i) = 0$ (summation is assumed to take place whenever the same index appears above and below).

The first advances in quantum electrodynamics did not occur until a general quantum field theory had been outlined by HEISEN-BERG and PAULI in March 1929. They based their approach on quantum mechanics by treating a field as a system of infinitely many degrees of freedom. In place of the co-ordinates q_k, they used the field quantities $Q(x)$ defined at a point x in three-dimensional space, such as $\psi(x)$, $\psi^*(x)$, $U(x)$ and $A(x)$. In place of the Lagrangian $L\,(q_k, \dot{q}_k)$, which led to the equations of motion via a variational calculation, they took the integral of a 'Lagrangian density'

$$\int L\left(Q_\alpha, \frac{\partial Q_\alpha}{\partial x^i}, \dot{Q}_\alpha\right) d\tau$$

which gave the field equations if the condition

$$\int L \, d\tau \, dt = \text{Min or Max}$$

was applied. The field quantities

$$P_\alpha(x) = \frac{\partial L}{\partial \dot{Q}_\alpha}$$

appeared in places of the canonical conjugate momenta

$$p_k = \frac{\partial L}{\partial \dot{q}_k}$$

and corresponding to the Hamiltonian $H = \sum p_k \dot{q}_k - L$ they formed a 'Hamiltonian density'

$$H = \sum P_\alpha(x) \dot{Q}_\alpha(x) - L$$

and a Hamiltonian $\int H \, d\tau$. The canonical equations that these gave were again the field equations. The transition to quantum theory was effected by means of commutation relations, the most fundamental of which were:

$$i\left[P_\alpha(x)Q_\beta(x') - Q_\beta(x')P_\alpha(x)\right] = \hbar \delta_{\alpha\beta} \delta(x - x').$$

The KLEIN-JORDAN commutation relations for the SCHROEDINGER field thus followed. The commutation relations for the field corresponding to the DIRAC equation of the electron were written with plus and minus signs by HEISENBERG and PAULI. This led to a certain difficulty in quantum electrodynamics as the quantity canonically conjugate to $U(x)$ became zero. The difficulty was disposed of by eliminating U and using various other tricks.

Rather more serious difficulties in quantum electrodynamics were later caused by the tendency of the expression for the energy of an electron to become infinite, this energy being derived from the interaction with the electrodynamic field. The same thing occurred in the case of the 'polarization of the vacuum' which was bound up with the possibility of the virtual creation of electron-positron pairs. These 'divergence problems' were eventually avoided rather than explained.

After the DIRAC equation had been established there was a persistent belief that this equation for the electron with one spinor as the field quantity represented the only possible relativistic quantum theory of matter and that therefore the spin $\hbar/2$ was the only possible one for an elementary particle. It was therefore a major advance when PAULI and Victor WEISSKOPF developed the quantum theory of a scalar field—that of the KLEIN-SCHROEDINGER-GORDON field—in 1934.[12] In the scalar theory it was possible for the density, already given by GORDON (cf. (4)) as

$$\rho \sim \frac{1}{2i}(\psi^*\dot\psi - \dot\psi^*\psi),$$

to take either sign. In DIRAC's theory, on the other hand,

$$\rho \sim \psi^*\psi$$

where ψ is a four-component quantity, had only one sign. The energy density was positive in the scalar theory, while according to DIRAC it could take either sign. There were indeed transitions from states of positive energy to ones of negative energy. In the scalar theory the number of particles was not necessarily constant. There was no conservation law for the number of particles, only a law for the conservation of electric charge. In the DIRAC theory the number of electrons had been constant before the theory had been extended to the 'hole theory'. That was after all why it fitted into quantum mechanics. After the theory had been extended, the appearance of an electron and a hole had been interpreted as a change in the number of particles, while electric charge was conserved. The scalar theory saw the creation or annihilation of particle-antiparticle pairs as the possible effect of an electromagnetic field without the aid of a hole theory. It was thus seen that the creation and annihilation of pairs of particles was a common feature of all relativistic quantum field theories. In Figure 23 the regions $E_{\mathrm{kin}} > mc^2$ and $E_{\mathrm{kin}} < -mc^2$ now denote regions where the charge has different signs. Figures 24 and 25 are also valid in this context. This meant that a distinction had to be drawn between two different types of field. There was one type with a definite (i.e., only one possible) sign for charge density, and with indefinite energy. Only in the case of FERMI statistics did this have an acceptable physical interpretation: the spin had the magnitude $\hbar/2$. The second type had an indefinite charge but a definite energy, which was possible also for BOSE statistics. The scalar field had zero spin. PAULI later proved that particles and fields with half-integer spin had to satisfy FERMI statistics and those with integral spin followed BOSE statistics.[13]

Coupling of Fields or Particles

The first interaction between various fields or particles to be recognized was the electromagnetic coupling. This was dealt with by introducing the electromagnetic potentials into the non-relativistic, the relativistic scalar or spinor fields. The coupling was later to be represented by diagrams of the type shown in Figure 28. Such a diagram can be read in six different directions with respect to time, four of them essentially different. It shows emission of a quantum of light, absorption of a quantum of light, creation of an electron-positron pair (other particles being required to satisfy the laws of energy and momentum), and the annihilation of an electron-positron pair.

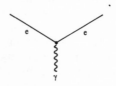

FIGURE 28: γee COUPLING

The discovery of the neutron by CHADWICK in February 1932 made it possible to formulate a new conception of the structure of the atomic nucleus and of the nature of β decay. There had hitherto been one or two difficulties associated with β decay. Although the initial and final states of the nucleus had definite values of the energy, β radiation displayed a continuum. Furthermore, nuclei of even mass number had integral spin and those of odd mass number had half-integral spin. The electrons which had been assumed to be in the nucleus together with the protons thus made no contribution whatsoever to the spin. The magnetic moments of the nuclei were also too small to include a contribution from the magnetic moment of electrons. And finally nuclei of even mass number satisfied BOSE statistics, while those of odd mass number followed FERMI statistics. The electrons in the nucleus had thus apparently lost not only their spin but also their statistical value. In order to get round these difficulties PAULI conceived of a new uncharged particle. It was therefore very difficult to observe: it also had only a small or negligible mass, half-integer spin, and it satisfied FERMI statistics; it was emitted from the nucleus in the course of β decay. The particle was later to be called the neutrino.[14]

It was assumed soon after the discovery of the neutron that protons and neutrons must be the constituents of the nucleus. This made it possible to understand the laws governing the spin and statistics of nuclei. β particles arose as a result of the transformation of nuclei, just as quanta of light were generated as a result of the transition of atoms to different energy states. According to HEISENBERG (June 1932) the possibility of the emission of an electron when a neutron was transformed into a proton led to forces between the nuclear particles, just as the possibility of the generation or absorption of electrons could lead to Coulomb forces. FERMI then went on in 1934 to give a quantitative theory of β decay, postulating for the fields of the proton, the neutron, the electron and the neutrino an interaction that gave the terms

$$np^* e^* v^* \qquad n^* p e v$$

in the Hamiltonian, where n was the annihilation operator of a neutron state, p^*, e^* and v^* were the creation operators for proton, electron and neutrino states. n and p were treated non-relativistically, and e and v were given the four DIRAC components. FERMI was able to deduce the form of the energy spectrum of β decay and a criterion for a very small or vanishing mass of the neutrino. TAMM and IVANENKO suggested that the FERMI coupling led either to nuclear forces that were far too weak or to distances between the nuclear particles that were far too small.[15]

In 1935, Hideki YUKAWA showed that the empirical value for the extent of the region over which the nuclear forces were effective could be explained by the use of a static matter field. Corresponding to the field equation

$$\Delta\psi - \kappa^2 \psi = 0$$

with the solution

$$\psi \sim \frac{e^{-\kappa r}}{r}$$

there must be forces with an effect over a distance $1/\kappa$ between the sources of the matter field. By the formula $\hbar\kappa = mc$ he took a mass of approximately 200 electron-masses for the masses of the particles that corresponded to the matter distribution, in order to explain the empirical values of the effective distance. If we give these particles the name they eventually received, π-mesons, this means that YUKAWA introduced an (npπ) coupling, in which terms of the type

$$np^*\pi_-^* \qquad np^*\pi_+ \qquad n^*p\pi_- \qquad n^*p\pi_+^*$$

arose, which explained the forces within the nucleus. He also introduced a (πev) coupling for β decay, with the terms

$$\pi_+ e_+^* v^* \qquad \pi_- e_-^* v.$$

Because the nuclear forces depended on the spin of the nuclear particles, YUKAWA and his collaborators (1938) had to assume a vector field for the meson. A field of this type had already been discussed by PROCA in 1936 (without the coupling of nuclear particles).[16] For the four vector components equations held (without coupling) which were analogous to MAXWELL's equations:

$$\frac{\partial U_k}{\partial x^i} - \frac{\partial U_i}{\partial x^k} = \kappa F_{ik}$$

$$\frac{\partial F^{ik}}{\partial x^k} = \kappa U^i$$

from which it followed that

$$\frac{\partial U^k}{\partial x^k} = 0$$

and that

$$\square\, U^k - \kappa^2 U^k = 0.$$

($\square$ is the four-dimensional analog of the operator Δ or ∇^2). Addition of the electromagnetic coupling led to additional terms in the wave equation which corresponded to a spin $\hbar$ and an associated magnetic moment of the particles that had been assigned to the field.

Thus, during the thirties, a general concept of the matter field had been developed. As it was possible to write the scalar field theory without coupling in the form

$$\frac{\partial \psi}{\partial x^i} = \kappa F_i$$

$$\frac{\partial F^i}{\partial x^i} = \kappa \psi$$

$$\square \psi - \kappa^2 \psi = 0$$

and VAN DER WAERDEN's spinor analysis allowed of a similar description of the DIRAC theory it was possible to regard equations of the type

$$\text{Der } \psi = \kappa \chi$$

$$\text{Der } \chi = \kappa \psi$$

as the basis of an uncoupled field theory. The field quantities ψ and χ satisfied commutation relations that followed from the canonical system. ψ and χ were scalars, spinors, vectors or tensors and Der denotes covariant differentiation with respect to the co-ordinates. A scalar ψ gave spin-free particles, a spinor ψ gave the spin $\hbar/2$, a vector ψ the spin $\hbar$, etc. The introduction of the electromagnetic four-potential in the combination

$$\frac{\hbar}{i} \frac{\partial}{\partial x^k} - e A_k$$

led (except in the scalar theory) not to the simple wave equation but rather to additional terms that described the electromagnetic behaviour of the spin. LORENTZ- and gauge-invariance had to be taken into account for the coupling of various kinds of matter. Many of these types of coupling turned out to be suitable for the description of the transformations of elementary particles.

[1] O. KLEIN, Z. Phys. **37**, 895 (1926), **41**, 407 (1927)
   E. SCHROEDINGER, Ann. Phys. **81**, 109 (1926)
   W. GORDON, Z. Phys. **40**, 117 (1926)
   H. WEYL, Gruppentheorie und Quantenmechanik, Leipzig 1931
[2] P. A. M. DIRAC, Proc. Roy. Soc. **117**, 610, **118**, 35 (1928)
[3] W. GORDON, Z. Phys. **48**, 11 (1928), **50**, 630 (1928)
   C. G. DARWIN, Proc. Roy. Soc. **118**, 654 (1928)
   O. KLEIN and Y. NISHINA, Z. Phys. **52**, 853 (1928)
[4] B. L. VAN DER WAERDEN, Nachr. Akad. Gött. **1929**, 100
[5] O. KLEIN, Z. Phys. **53**, 157 (1929)

[6] P. A. M. Dirac, Proc. Roy. Soc. **126**, 360 (1930)

[7] J. R. Oppenheimer, Phys. Rev. **35**, 562, 939 (1930)
P. A. M. Dirac, Proc. Cambr. Philos. Soc. **26**, 361 (1930), Proc. Roy. Soc. **133**, 60 (1931)
I. Tamm, Z. Phys. **62**, 545 (1930)

[8] W. Pauli, Die allgemeinen Prinzipien der Wellenmechanik, Hdb. d. Phys. 2. Aufl. **24**/1, 83 (1933)

[9] P. A. M. Dirac, Proc. Roy. Soc. **114**, 243, 710 (1927)

[10] P. Jordan and W. Pauli, Z. Phys. **47**, 151 (1928)

[11] W. Heisenberg and W. Pauli, Z. Phys. **56**, 1 (1929)

[12] W. Pauli and V. Weisskopf, Helv. Phys. Acta **7**, 709 (1934)

[13] W. Pauli, Phys. Rev. **58**, 716 (1940)

[14] cf. C. S. Wu in Pauli Memorial Volume (1960)

[15] W. Heisenberg, Z. Phys. **77**, 1 (1932)
E. Fermi, Z. Phys. **88**, 161 (1934)
I. Tamm, Nature **133**, 981 (1934)
D. Ivanenko, Nature **133**, 981 (1934)

[16] H. Yukawa, Proc. Phys.-Math. Soc. Jap. **17**, 48 (1935)
H. Yukawa, S. Sakata and M. Taketani, ibid. **20**, 319 (1938)
A. Proca, Jn. de phys. **7**, 341 (1936)

APPENDIX: AN OUTLINE OF
QUANTUM MECHANICS

Survey

THIS outline of quantum mechanics—i.e., the mechanics of the atomic region—is intended to proceed as far as possible in historical order. Quantum mechanics may be regarded as a non-intuitive modification of classical particle mechanics or as a non-intuitive modification of an intuitive field theory of matter.

We can maintain the link with classical mechanics because we can still define the basic concepts of position, momentum, time and energy. We must modify these concepts because it is impossible to make simultaneous precise measurements of, for example, position and momentum. Physical quantities can therefore not always be represented by ordinary numbers but may require more general mathematical models. Just as in classical mechanics it is possible to reveal the essentials of quantum mechanics even for only one degree of freedom: the theorem that lies at the heart of this form of quantum mechanics is the commutation relation

$$i(pq - qp) = \hbar \tag{1}$$

for co-ordinate q and momentum p. It is necessary to base this on experimental evidence and a very fruitful empirical theorem for this purpose is the spectral combination principle

$$v = F(n + \tau) - F(n).$$

We really have only a little more to do than is required by the analysis of this principle if all we want to do is examine the 'old quantum mechanics' or the 'quantum mechanics of the correspondence principle'.

We must build our intuitive field theory for matter on the experimental evidence that a stream of matter with uniform properties and moving uniformly displays wave properties and the fact that the wave number depends on the electric potential in the case of electrically charged matter of a given frequency. The action of matter upon itself is not easy to take into account. We shall therefore first consider the limiting case of matter that is so thinly distributed that we may neglect its action upon itself. The quantum modification of this field theory is required by the fact that in reality matter also has particle properties. It is then not possible to obtain the limiting case of arbitrarily thinly-distributed matter. But as a particle does not exert any force upon itself we may develop a quantum theory for a particle from the intuitive theory for thinly-distributed matter. It agrees with that derived from particle mechanics.

In this appendix we shall therefore consider classical mechanics, old quantum mechanics, the commutation relation, its satisfaction in terms of suitable mathematical models (matrices and differential operators), the intuitive field theory of matter and its modification which leads to the SCHROEDINGER equation.

Mechanics

Quite often a physical system is treated as a system of mass particles. A particular system of this type is then defined in terms of the forces F_k, which appear in the equations of motion

$$m_k \ddot{x}_k = F_k(x_1, x_2 \ldots)$$

for the co-ordinates $x_k(x_1, x_2, x_3$ for the first particle, x_4, x_5, x_6 for the second, etc.). It is therefore also possible to write this in the form

$$m_k \dot{x}_k = p_k$$

$$\dot{p}_k = F_k(x_1, x_2 \ldots)$$

and to transform it into:

$$\dot{q}_k = \frac{\partial H}{\partial p_k} \qquad \dot{p}_k = -\frac{\partial H}{\partial q_k} \tag{2}$$

where H is treated as a function of the co-ordinates q_k (here equal to x_k) and the momenta p_k. A particle moving along a straight line, for example an oscillator, is then characterized by

$$H = \frac{p^2}{2m} + V(x)$$

where $V(x)$ has for example the form shown in Figure 29. It is possible to describe by a system of this kind even rigid bodies or other systems with rigid conditions. In place of the x_k, we then have other generalized co-ordinates q_k, for example angles. These then have canonically conjugate generalized momenta p_k which may, for example, be angular momenta. Thus a 'rotator' is characterized in terms of

$$H = \frac{P^2}{2I} + V(\varphi)$$

where P is the angular momentum and I the moment of inertia. The 'kinematics' of a system are described by giving the numerical value and the physical explanation of the q_k and p_k. Its 'dynamics' are described by giving the Hamiltonian $H(q_1, q_2, \dots p_1, p_2 \dots)$, and its development with respect to time is governed by the 'canonical equations of motion' (2). When the values of the variables q_k and p_k are known at some initial point of time its future development is completely determined. In the cases that interest us, the appropriate value of H gives the energy. If there is no external influence (i.e., H does not depend explicitly on time) the energy is constant with respect to time.

In 'classical mechanics' it is assumed either tacitly or explicitly that all the q_k and the p_k may be obtained separately. The q_k and p_k are quantities with a definite numerical value. Thus a classical system is 'determined' in the sense given. *In quantum theory we must regard the q_k and p_k as known only within the limits of quantum-theoretical uncertainty.* Thus a quantum-mechanical system is not strictly causally determined. It is possible to make only probabilistic assertions about the future.

Statistical Mechanics

In thermodynamics we describe a physical system by means of macroscopic variables (e.g. pressure, density and temperature). In statistical mechanics we extend the description by the introduction of microscopic (mechanical) variables, for example the q_k and p_k of atoms. We explain the thermodynamic quantities temperature T and entropy S by finding those components of a more refined description which behave like these quantities. We then derive theorems relating to the macroscopic variables. Then, according to whether this is done in terms of classical mechanics or quantum mechanics, we speak of classical or quantum statistics. We thus have the following arrangement:

	Axiom of Certainty	Quantum-theoretical Uncertainty
Mechanical Description	Classical Mechanics	Quantum Mechanics
Macroscopic Description	Classical Statistics	Quantum Statistics

There are a number of theorems in statistical physics which do not actually depend on the form of physics (classical or quantum) being used. Among them are the definitions of entropy and temperature. Thus we define

$$S = k \ln W \tag{3}$$

where W is the number of events (the number of micro-states) needed to realize a macro-state. For sufficiently large systems one of the macro-states has an overwhelming maximum of W. It is the expected equilibrium state. Thus, for example, the state of uniform density of a gas is realized by far, far more events than a state of noticeably non-uniform density. Further, a system has temperature T if it may be regarded as one term of a canonical ensemble, i.e., as a term of an imagined ensemble of N systems with identical mechanical quantities which may exchange energy and in which a particular micro-state with energy E_l occurs with frequency

$$\sim e^{-\frac{E_l}{kT}} = e^{-\beta E_l}. \tag{4}$$

The ensemble (4) is the most probable of all the possible ensembles with given total energy $N\bar{E}$. The temperature is thus a property of an equilibrium state. However, an entropy occurs also in non-equilibrium states. The transition to equilibrium causes the entropy to increase. At a given temperature T the average value $\bar{E}$ of the energy is given by

$$\bar{E} = \frac{\Sigma E_l e^{-\beta E_l}}{\Sigma e^{-\beta E_l}} = -\frac{d}{d\beta} \ln \Sigma e^{-\beta E_l} \tag{5}$$

where the summation is carried out over all the micro-states l, and not, for example, over all the different energies.

Classical statistics diverges from quantum statistics in respect

Appendix: An Outline of Quantum Mechanics   217

of the way in which events are counted. In classical mechanics the space of the canonical variables, 'phase space', is an ordinary multi-dimensional geometrical space, and there are good reasons for taking the number W of events for a macro-state as the volume Φ of the region of phase space corresponding to the macro-state

$$W \sim \Phi = \iint \ldots dq_1 \, dq_2 \ldots dp_1 \, dp_2 \ldots$$

The unit of phase extension, i.e., the ratio of Φ to W, remains arbitrary. As the transition from one unit Ω for $dp \, dq$ to a new unit ω causes the quantity W to be multiplied by $(\Omega/\omega)^N$ and thus increases the entropy by an amount $kN \ln (\Omega/\omega)$ the entropy in classical statistics includes an arbitrary additive constant. In quantum statistics the space of the canonical variables cannot be treated as an ordinary geometrical space because of the quantum-mechanical uncertainty. It is, however, possible to define individual quantum states $l = 1, 2, \ldots$ which each count as a single event.

In classical statistics there are equipartition theorems for the energy. Thus the kinetic energy consists of the contributions $p^2/2m$ from the individual degrees of freedom, and the average value of these quantities in thermal equilibrium is given by:

$$\frac{\overline{p^2}}{2m} = -\frac{d}{d\beta} \ln \int e^{-\frac{\beta p^2}{2m}} \, dp = \frac{1}{2\beta} = \frac{1}{2} kT.$$

Each degree of freedom contributes on average the amount $kT/2$ to the kinetic energy. For the potential energy there is in general no simple rule. However, for harmonic oscillators, the potential energy is proportional to q^2 and this likewise gives $kT/2$ for the average value. If we idealize a rigid body composed of N atoms by means of $3N$ harmonic oscillators its energy comes out to be $3NkT$, its atomic heat to be $3R$, as given by the DULONG-PETIT rule.

The Anharmonic Oscillator

We shall later explain the basic theories of quantum theory in terms of a simple mechanical system, a system with one degree of freedom, with canonical variables x and p, executing periodic motion. This is thus a system with Hamiltonian

$$H(p,x) = \frac{p^2}{2m} + V(x) \tag{6}$$

where $V(x)$ behaves, say, as in Figure 29. Let us first investigate this system in terms of classical mechanics. The possible paths are represented by points in the x, p-plane which travel along the curves

$$p(E,x) = \pm \sqrt{2m[E - V(x)]}$$

thus enclosing the areas

$$\Phi(E) = \oint p(E,x)\,\mathrm{d}x$$

(Figure 29). The motion comprises the frequencies

$$\nu = \tau \nu_1(E)$$

of which $\nu_1(E)$ can be calculated from $V(x)$:

$$\frac{1}{\nu_1} = \oint \mathrm{d}t = \oint \frac{\mathrm{d}x}{v(E,x)} = \oint \frac{\partial p(E,x)}{\partial E}\,\mathrm{d}x = \frac{\mathrm{d}}{\mathrm{d}E}\oint p(E,x)\,\mathrm{d}x = \frac{\mathrm{d}\Phi(E)}{\mathrm{d}E}.$$

At the third 'equals' sign we must use the equation $\mathrm{d}E = v\,\mathrm{d}p$, which holds for all values of x. At the fourth we have to recall that the limits of integration depend on E but contribute nothing to the derivative with respect to E in this case, as p disappears at the limits of integration.

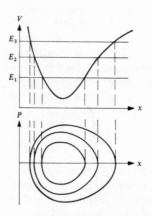

FIGURE 29: POTENTIAL AND PHASE REGION OF AN ANHARMONIC OSCILLATOR

FIGURE 30: ADIABATIC INVARIANCE

From the function $V(x)$ we obtain the function

$$\Phi(E) = \oint p\,dx = \oint 2E_{kin}\,dt = \frac{2\bar{E}_{kin}}{v} \tag{7}$$

and the frequencies

$$v = \tau \frac{dE}{d\Phi}. \tag{8}$$

If Φ contains some other parameter a it naturally follows that Φ and v_1 depend on E and a. It can be shown that for slow—adiabatic—change of a the motion goes over into another with an identical value of Φ, and thus that the quantity Φ does not change during the motion. $\Phi(a,E)$ is adiabatically invariant. For a harmonic oscillator v is independent of E, and thus

$$\Phi = \frac{E}{v}.$$

We can illustrate the adiabatic invariance of this quantity by means of a simple pendulum. Let us change the length l of the pendulum by pulling the string up through a narrow hole (see Figure 30). The tension in the string is made up of components of the weight and the centripetal force, so that the work done is equal to

$$\delta A = -(mg\cos\varphi + ml\dot{\varphi}^2)\delta l.$$

where φ is the angular displacement. During the adiabatic change that we caused by raising the string a large number of oscillations occur without l changing appreciably, so that we may write the equation in terms of average values.

$$\delta A = -(mg\overline{\cos\varphi} + ml\overline{\dot{\varphi}^2})\delta l.$$

If we now separate the energy increase δA into the increase in external and internal energy

$$\delta A = -mg\,\delta l + \delta E$$

we have, for the internal energy,

$$\delta E = [mg(1 - \overline{\cos\varphi}) - ml\,\overline{\dot\varphi^2}]\delta l = (\overline{E_{\text{pot}}} - 2\overline{E_{\text{kin}}})\frac{\delta l}{l}.$$

So long as the oscillation is harmonic, we have $\overline{E_{\text{pot}}} = \overline{E_{\text{kin}}}$, and thus

$$\frac{\delta E}{E} = -\frac{1}{2}\frac{\delta l}{l}$$

$$E\sqrt{l} \sim \frac{E}{v} = \text{const.}$$

If we pull the string slowly through the hole, the oscillation energy changes in proportion to the frequency.

Let us now consider the relationship between $V(x)$, $v_1(E)$ and $\Phi(E)$ by looking at examples. Let us choose those for which the fundamental frequency v_1 is constant, those for which v_1 increases with E and those for which v_1 decreases with E. For the harmonic oscillator $\tau = 1$ and v is constant. It follows from (8) that E equals $v\Phi$, if we calculate E with respect to the lowest value of V. E/v is, as we have already said, adiabatically invariant. For motion within a 'box'

$$V = 0 \qquad 0 < x < a$$

and elastic reflection at $x = 0$ and $x = a$ we have

$$p = \pm 2mav_1 \qquad E = 2ma^2v_1^2.$$

Furthermore

$$\Phi = 4ma^2v_1$$

from which it is possible to deduce equation (8). Finally, we have

$$E = \frac{\Phi^2}{8ma^2}. \tag{9}$$

Related to this system is that of the free 'rotator' with canonical variables φ and P, I being the moment of inertia:

$$P = Iv \qquad E = \frac{I}{2}v^2.$$

We further have that

$$\Phi = 2\pi P$$

from which we can see (8), and finally that

$$E = \frac{\Phi^2}{8\pi^2 I}. \tag{10}$$

For a potential energy $V \sim -1/x$ $(x > 0)$ and perfectly elastic reflection at $x = 0$ we have

$$v_1 \sim |E|^{3/2}.$$

This is the limiting case of Keplerian motion in a potential $V \sim -1/r$. Applying (8) it follows that $d\Phi/dE \sim |E|^{-3/2}$ and that

$$E \sim -\frac{1}{\Phi^2}. \tag{11}$$

The three examples all fall under the general description $v_1 = aE^r$, from which it follows that

$$E = [a(1-r)\Phi]^{\frac{1}{1-r}}. \tag{12}$$

$r = 0$, $r = \frac{1}{2}$ and $r = \frac{3}{2}$ correspond to the harmonic oscillator, the rotator and the potential $\sim -1/x$ respectively and lead to $E \sim \Phi$, $E \sim \Phi^2$, $E \sim -1/\Phi^2$.

The Correspondence Principle

In the atomic domain, classical mechanics no longer holds. We now look for a physics of the atom, ideally that which involves the least possible modification of classical mechanics. On the basis of the combination principle for spectra,

$$v = F(n + \tau) - F(n)$$

and of the relationship between the frequency and the transmitted energy for short-wave radiation, we regard it as an empirically certain starting point that an anharmonic oscillator may assume only discrete values of the energy $E(n)$ and that a frequency of emission or absorption is connected with the transition from one energy $E(n)$ to another in this sequence by the formula

$$v = \frac{E(n + \tau) - E(n)}{h}. \tag{13}$$

222 *The History of Quantum Theory*

The question then arises: is it possible to calculate the values $E(n)$ if we know $V(x)$? Our guide for this is the limiting transition from atomic physics—quantum mechanics—to classical physics for sufficiently large values of the energy. If we wish to proceed very carefully we must determine the $E(n)$ in such a way as to make the quantum frequencies (13) agree as well as possible with the classical frequencies

$$\nu = \tau \nu_1 = \tau \frac{dE}{d\Phi}. \tag{14}$$

Figure 31 is intended to illustrate this scheme: (a) denotes the values of $E(n)$ that remain to be determined. The distances between the lines then give, for each n, the emission frequencies to the left and the absorption frequencies to the right. (b) shows classical frequencies for $n = 1$, while (c) shows them for $n = 8$. The problem is now one of arranging the $E(n)$ in such a way that the classical frequencies correspond as closely as possible to the differences between them. This can only be done exactly if the classical frequency ν_1 does not depend on E. But it can be done approximately if $\Phi(E)$ is a fairly smooth function.

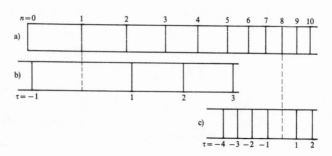

FIGURE 31: THE CORRESPONDENCE PRINCIPLE

The quantum frequency (13) is roughly

$$\nu = \tau \frac{dE}{h\,dn}$$

and this is the classical frequency (14) if we put $d\Phi = h\,dn$, i.e., we admit into quantum theory only the values of $E(n)$ for which we have $\Delta\Phi = h$. We must thus put

$$\Phi = hn$$

or

$$\Phi = h(n+\alpha)$$

and we shall take $\alpha = 0$ for the time being. In doing this we allow the classical formula for $\Phi(E)$ to be upheld. The agreement between the quantum frequencies and the classical ones is the better, the smoother $\Phi(E)$ is. We may reasonably expect it to hold particularly well for large values of E.

Quantum theory allows only the states for which

$$\Phi = \oint p\,\mathrm{d}x = hn.$$

If we know the classical formula for $v_1(E)$ we can calculate the permitted energies $E(n)$ from the formula

$$\int \frac{\mathrm{d}E}{v_1(E)} = hn \tag{16}$$

(the HASENÖHRL quantum condition).

Let us take another look at these last few examples. For the harmonic oscillator $\Phi = E/v$ leads to $E = hvn$. For a box (9) gives:

$$E = \frac{h^2 n^2}{8ma^2}. \tag{17}$$

Emission and absorption frequencies are slightly different:

$$v(n,n-1) = \frac{h}{4ma^2}(n-\tfrac{1}{2})$$

$$v(n+1,n) = \frac{h}{4ma^2}(n+\tfrac{1}{2}).$$

For a rotator we have

$$P = \frac{h}{2\pi}n$$

for which we may also write

$$P = \hbar n \tag{18}$$

and

$$E = \frac{\hbar^2 n^2}{2I}. \tag{19}$$

For a potential $V = c/x$, as for example in the case of Kepler-type motion, we have

$$v_1^2 = \frac{2|E|^3}{\pi^2 m e^2}.$$

Equation (16) thus leads to

$$E = -\frac{2\pi^2 m c^2}{h^2 n^2}. \tag{20}$$

We started out from a correspondence principle for the frequencies. We could instead have begun with the empirical result that the quantity W had a definite value in the equation $S = k \ln W$. This suggests the introduction of a definite unit h for the region of phase space Φ which gives the number of events to within an arbitrary factor in classical statistics. We thus in effect count a region of phase space $\Delta\Phi = h$ as a single event, by admitting only those states with $\Phi = hn$. For large values of n quantum statistics then becomes classical statistics—a correspondence principle for statistics. We can proceed as follows. The classical equation

$$v = \tau \frac{dE}{d\Phi}$$

has a meaning, for discrete values of Φ, only if Φ is a fairly smooth function of E. It must be strengthened into an equation that has a meaning for all $\Phi = hn$ and which tends to the classical equation for smooth functions $\Phi(E)$. And (13) would seem to be the obvious choice.

To enable us to put $\Phi = hn$, we must make one further assumption. For a slow change of one parameter, permitted states must transform into permitted states. If the change is a quick one, we get 'quantum jumps'. Φ must therefore be an adiabatic invariant. This is the case in classical mechanics.

The formula $\Phi = hn$ for the calculation of the energy values $E(n)$ cannot be true in general. In the calculation the potential energy $V(x)$ is used only in the region $V \leq E$—the region of the classical motion. However, among the properties of the stationary states there are their radiation frequencies $v(n + \tau, n)$ which are indeed connected with its behaviour in the region $V > E$. A particularly striking case is that of a potential barrier (Figure 32).

If we represent the energy $E(n)$ as given by the phase integral in a diagram which has the height W of the barrier as its abscissa we obtain for $E > W$, in the case of a symmetrical potential, narrower intervals, and for $E < W$ further intervals which do not have a continuous transition for $E = W$ (Figure 32b). It would be more realistic to expect behaviour of the type shown in Figure 32c. In the light of the correspondence principle, the failure of $\Phi = hn$ is not really surprising. The method involves the replacement of a difference by a differential quotient. ΔE becomes dE/dn, which is a reasonable approximation for smooth functions $v(E)$ and $E(n)$. However, $v(E)$ is *not* a smooth function at the barrier.

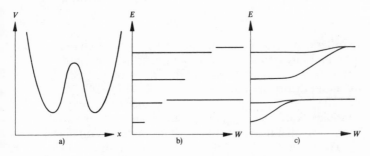

FIGURE 32: THE FAILURE OF $\Phi = hn$

The Correspondence Principle for several Degrees of Freedom

The extension of the quantum theory of the correspondence principle to take in several degrees of freedom is now of purely historical significance. We can write down the correspondence between the quantum frequencies and the classical ones for systems whose co-ordinates execute the 'multiply-periodic motion'

$$x = \sum C_{\tau_1 \tau_2} e^{2\pi i(\tau_1 v_1 + \tau_2 v_2 + \ldots)t}$$

thus comprising frequencies

$$v = \tau_1 v_1 + \tau_2 v_2 + \ldots \tag{21}$$

The quantum frequency

$$v = \frac{1}{h}\left[E(n_1 + \tau_1, n_2 + \tau_2 \ldots) - E(n_1, n_2 \ldots)\right] \tag{22}$$

corresponds to a classical frequency of this kind. The correspondence principle requires the quantities (21) and (22) to agree more closely with each other the smoother the function $E(n_1, n_2,...)$. Then, however, (22) becomes approximately

$$v = \tau_1 \frac{\partial E}{h \partial n_1} + \tau_2 \frac{\partial E}{h \partial n_2} + \cdots$$

If we can thus write the classical frequencies $v_1, v_2, ...$ in the form

$$v_k = \frac{\partial E}{\partial I_k} \qquad (23)$$

we can satisfy the correspondence principle by putting

$$I_k = h n_k. \qquad (24)$$

In order to provide a quantum description of a mechanical system with multiply-periodic motion we must therefore look for variables $I_1, I_2,...$ that satisfy (23). The motions permitted by quantum theory are then those determined by (24)

If the system is 'separable' (cf. Chapter 6) we may introduce canonical variables w_k and I_k—angle and action variables—in such a way that the Hamiltonian H depends only on the variables I_k. The quantities I_k and $\dot{w}_k = \partial H / \partial I_k$ are thus constants of the motion. For periodic motion the variables can be chosen in such a way that the $\dot{w}_k$ are the frequencies, and the equations (23) are thus satisfied.

Separation occurs when the conservation laws that follow from the invariance properties of the system, for example those of energy and angular momentum, provide as many constants of the motion as the number of degrees of freedom. For a smaller number of constants of the motion the motion can be described partially in terms of w_k and I_k.

For rotation invariance with respect to a point (for example in the case of a free atom) the size P of the angular momentum and its components P_z about an arbitrary axis are constants of the motion, but P_z does not appear in the energy. $2\pi P$ and $2\pi P_2$ are action variables. In quantum theory we therefore put

$$P = \hbar J \qquad J = 0, 1, 2...$$

It is possible to establish the selection rule $\Delta J = 0, \pm 1$.

If it is periodic, the classical motion of a particle in a central field $V(r)$, in general has two fundamental frequencies, to which there correspond two quantum numbers n and l. For two particular forms of $V(r)$ there is only one fundamental frequency. In the case of $V \sim -1/r$ (Keplerian motion) the quantum-theoretical energy depends on only one quantum number n. A more precise investigation shows that $E \sim -1/n^2$ with $n = l + n_r + 1$. For $V \sim r^2$ (the harmonic oscillator) we see that $E = \hbar\omega n$ with $n = l + 2n_r$.

Exposure of an atom to a weak magnetic field in the z direction causes a uniform precession about the z direction in addition to the other components of the motion of the electrons; the energy also depends on P_z. In quantum theory, the energy depends on m and we have to assume that $m \leq J$. For an atom in an external electric field the energy cannot depend on the sign of m but only on its magnitude, for reasons of symmetry.

If it is possible to treat parts of an atom as weakly coupled it is possible to apply the angular momentum laws approximately to each part separately and then add the angular momenta vectorially. To a large extent this is the basis of the classification of atomic spectra.

The Matrix Form of Quantum Mechanics

It is a basic fact of quantum theory that frequencies belong to two states. If we consider a mechanical system with a single degree of freedom there is a single series of quantum numbers, and we can denote the frequency by $\omega(n,l)$. The frequencies satisfy the combination principle

$$\omega(n,k) + \omega(k,l) = \omega(n,l) \tag{25}$$

if we put $\omega(l,n) = -\omega(n,l)$. A quantity whose classical analogue varies periodically with time, and which can therefore be written in the form

$$x(E,t) = \sum_{\tau} x_{\tau}(E)e^{i\tau\,\omega(E)t} \tag{26}$$

belongs, as does the frequency, to two states. In quantum theory we write a quantity of this type as a square array

$$x(n,l) = x_{nl}e^{i\,\omega(n,l)t}. \tag{27}$$

In the case of real quantities in classical mechanics we have $x_{-\tau} = x_\tau^*$. We therefore require the corresponding ensembles to satisfy:

$$x_{ln} = x_{nl}^*. \tag{28}$$

A quantity whose classical analogue corresponds to a single state, as energy does for example, is written as a diagonal array

$$E(n,l) = E_n \delta_{nl}.$$

It is with these ensembles that we calculate. As the multiplication of classical quantities represented by Fourier series is carried out according to the law

$$xy = \sum_\rho x_\rho e^{i\rho\omega t} \sum_{\tau-\rho} y_{\tau-\rho} e^{i(\tau-\rho)\omega t} = \sum_\tau z_\tau e^{i\tau\omega t} \qquad \sum x_\rho y_{\tau-\rho} = z_\tau$$

where it is important to note that

$$\rho\omega + (\tau - \rho)\omega = \tau\omega$$

we multiply ensembles of the type (27) according to the rule:

$$\sum_k x(n,k) y(k,l) = z(n,l) \qquad \sum_k x_{nk} y_{kl} = z_{nl}. \tag{29}$$

We have of course used the combination principle (25). We thus treat these ensembles as matrices. The matrix with the element $x(l,n)^*$ in the n, l position is called the matrix adjoint to $x(n,l)$ and is written as x^* for short. Matrices of the type (28) are self-adjoint, so that $x^* = x$. They are also called Hermitian matrices. Even if x and y are Hermitian, xy need not be. However, $xy + yx$ and $i(xy - yx)$ are certainly Hermitian. If x and y are Hermitian we have

$$\xi(n,l) = x(n,l) + i y(n,l) = x(l,n)^* + i y(l,n)^* = (x - i y)(l,n)^*.$$

If

$$x + i y = \xi$$

then clearly

$$x - i y = \xi^*.$$

Hermitian matrices correspond to real numbers; ξ and ξ^* behave to a large extent like conjugate complex numbers.

In quantum dynamics we require equations of motion for the matrices corresponding to the physical quantities. By (27) we have to put

$$\dot{x}(n,l) = \omega(n,l) x(n,l). \tag{30}$$

If we now draw on the basic experimental fact that

$$\hbar\omega(n,l) = E_n - E_l$$

we can replace (30) by

$$\hbar\dot{x}(n,l) = i[E(n,n)x(n,l) - x(n,l)E(l,l)] \qquad \hbar\dot{x} = i(Ex - xE).$$

If we characterize the mechanical system in terms of a Hamiltonian $H(p,q)$, the equation of motion for any quantity x must be:

$$\hbar\dot{x} = i(Hx - xH). \tag{31}$$

It follows that $\dot{H} = 0$ (we are considering only conservative systems).

The equations of motion (31), which must include those for the canonical variables

$$\hbar\dot{q} = i(Hq - qH) \qquad \hbar\dot{p} = i(Hp - pH)$$

are still not sufficient. We need a correspondence principle which gives a close connection between these equations and the classical canonical equations of motion

$$\dot{q} = \frac{\partial H}{\partial p} \qquad \dot{p} = -\frac{\partial H}{\partial q}.$$

We obtain a close connection of this kind if we put

$$\hbar\frac{\partial F}{\partial p} = i(Fq - qF) \qquad \hbar\frac{\partial F}{\partial q} = i(pF - Fp) \tag{32}$$

$F(p,q)$ for any function. If we first take F as either p or q itself we see that

$$i(pq - qp) = \hbar. \tag{33}$$

It follows from this that (32) holds for any function F that can be formed by multiplication and addition from p and q (if the equations hold for F_1 and F_2, they follow for $F_1 + F_2$ and F_1F_2). We regard (33) as a fundamental postulate of quantum mechanics. It follows from (33) that it is meaningless to give p and q simultaneous numerical values. It is thus impossible to make simultaneous precise measurements of canonically conjugate quantities. (33) does not cause any modification in classical mechanics for relationships between macroscopic quantities, where we have $pq \gg \hbar$.

The aim of quantum mechanics now consists in finding matrices for the canonical variables p and q which satisfy equation (33), whose rows and columns correspond to the states of the system, and for which the Hamiltonian is thus a diagonal matrix. As a transformation of the type SxS^{-1} of any quantity x, where S is a matrix, preserves equation (33), and as the 'unitary transformation' SxS^* ($SS^* = 1$) transforms a Hermitian matrix x into a Hermitian matrix, we may formulate the problem as follows: let us assign matrices to the variables p and q that satisfy the commutation relations and let us seek a unitary transformation SxS^* which transforms the Hamiltonian $H(p,q)$ into a diagonal matrix. The problem is similar to that of finding the principal axis transformation of a quadratic form.

The generalization to several degrees of freedom is immediate. The equation of motion (31) and thus also the equations of motion

$$\hbar \dot{q}_k = i(H q_k - q_k H) \qquad \hbar \dot{p}_k = i(H p_k - p_k H)$$

correspond to the classical canonical equation

$$\dot{q}_k = \frac{\partial H}{\partial p_k} \qquad \dot{p}_k = -\frac{\partial H}{\partial q_k}$$

provided we may put

$$\hbar \frac{\partial F}{\partial p_k} = i(F q_k - q_k F) \qquad \hbar \frac{\partial F}{\partial q_k} = i(p_k F - F q_k)$$

which is the case if

$$\left.\begin{array}{c} i(p_k q_l - q_l p_k) = \hbar \delta_{kl} \\[4pt] p_k p_l - p_l p_k = 0 \\[4pt] q_k q_l - q_l q_k = 0 \end{array}\right\} \qquad (34)$$

The canonical transformations preserve the form SxS^*, where $SS^* = 1$. The commutation relations are satisfied by treating the q_k as numbers and replacing the p_k by the differential operators $\hbar \partial/i \partial q_k$; this follows from the equations

$$\left(\frac{\partial}{\partial x} x - x \frac{\partial}{\partial x} \right) F(x, y ...) = F(x, y ...)$$

$$\left(\frac{\partial}{\partial x} \frac{\partial}{\partial y} - \frac{\partial}{\partial y} \frac{\partial}{\partial x} \right) F(x, y ...) = 0.$$

The classical equation

$$H(p_1, p_2 \ldots q_1, q_2 \ldots) - E = 0$$

thus leads to a partial differential equation, due to SCHROE-DINGER:

$$H\left(\frac{\hbar}{i}\frac{\partial}{\partial q_1}, \frac{\hbar}{i}\frac{\partial}{\partial q_2}\ldots q_1, q_2\ldots\right)\psi(q_1, q_2\ldots) - E\psi(q_1, q_2\ldots) = 0. \quad (35)$$

For a single particle this can be written as:

$$-\frac{\hbar^2}{2m}\Delta\psi + (V - E)\psi = 0. \quad (36)$$

The Separate Wave and Particle Aspects

The flow of matter in a cathode ray displays wave properties. The wave number k, which can be measured by interference, is proportional to the velocity of the matter flux:

$$v = \frac{k}{\sigma} \quad (37)$$

provided that the velocities involved are not too high. A stream velocity v is measurable only if the matter is not perfectly homogeneous. We must interpret it is as the group-velocity of a wave packet, which is related to frequency and wave number (per 2π units of length) by the equation $v = d\omega/dk$, while for phase velocity we have $u = \omega/k$. For the matter which forms the cathode ray we thus have

$$\frac{d\omega}{dk} = \frac{k}{\sigma} \qquad \omega = \frac{k^2}{2\sigma} + \omega_0.$$

The group-velocity $d\omega/dk$ of a wave packet is defined the more precisely, the shorter the range of k of the harmonic waves which it comprises. The wave packet must then be very long.

We thus interpret the limiting case of a homogeneous flow of unit velocity as a harmonic wave

$$\psi \sim e^{-i\omega t + ikx} \qquad \omega = \frac{k^2}{2\sigma} + \omega_0. \quad (38)$$

This matter wave is essentially different from the light wave for which $\omega = ck$ and $u = v = c$. The fundamental difference is that light has the velocity c, while matter can take any velocity v such that $0 \leq v < c$. (We shall be dealing only with the non-relativistic theory of matter, and it is only for this theory that (37) holds.)

We may regard a wave of the type (38) as a special case of a field. In the case of light the harmonic wave with $\omega^2 = c^2 k^2$ is a solution of the field or wave equation

$$\Delta \psi - \frac{1}{c^2} \ddot{\psi} = 0.$$

For matter we can derive $\omega = k^2 / 2\sigma$ (we shall put ω_0 equal to zero here), using the equation

$$-\frac{1}{2\sigma} \Delta \psi - i \dot{\psi} = 0. \tag{39}$$

Homogeneous matter in free space satisfies equation (39). Charged matter exerts forces upon itself, neglected in (39). Now let us consider electrically charged matter in an electric field with potential $U(x)$. For particles of mass m and electric charge e we have:

$$\frac{v^2}{2} + \frac{e}{m} U = \text{const.}$$

The ratio e/m is meaningful even without the concept of particles. If we denote it by ζ/σ we have an equation

$$\frac{k^2}{2\sigma} + \zeta U = \text{const.} \tag{40}$$

connecting group velocity v and electric potential. Experience shows that ζ has the same magnitude for cathode rays, proton radiation, α radiation, etc. In the case of cathode rays $\zeta < 0$, while for proton and α radiation $\zeta > 0$. σ is much larger for proton and α-radiation than for cathode rays. If a wave of unit frequency leaves a field-free region for another field-free region we must assume, by (38) and (40), that

$$\frac{k^2}{2\sigma} + \zeta U = \omega. \tag{41}$$

We may certainly ignore ω_0 as U is defined only to within an arbitrary additive constant. One wave equation which gives equation (41) for a harmonic wave is

$$-\frac{1}{2\sigma} \Delta \psi + \zeta U \psi - i \dot{\psi} = 0. \tag{42}$$

In regions of varying potential U (40) is meaningless but on the other hand (42) does have a meaning.

Homogeneous matter is represented by a field ψ with equation (41). The specific charge ζ/σ and the quantity σ which expresses the connection between wave number and group velocity can be measured experimentally.

It follows from (42) and its complex conjugate that

$$\frac{\partial}{\partial t}(\psi^*\psi) = \psi^*\dot{\psi} + \dot{\psi}^*\psi = \frac{i}{2\sigma}(\psi^*\Delta\psi - \psi\Delta\psi^*)$$

and thus that

$$\frac{\partial}{\partial t}(\psi^*\psi) + \operatorname{div}\left[\frac{1}{2i\sigma}(\psi^*\operatorname{grad}\psi - \psi\operatorname{grad}\psi^*)\right] = 0.$$

This is a conservation law for the amount of matter if we regard

$$\rho \sim \psi^*\psi$$

$$s \sim \frac{1}{2i\sigma}(\psi^*\operatorname{grad}\psi - \psi\operatorname{grad}\psi^*)$$

as the density of matter and of matter flux respectively. It is a law for electric charge if we take

$$\left.\begin{aligned}\rho &\sim \zeta\psi^*\psi \\[2mm] s &\sim \frac{\zeta}{2i\sigma}(\psi^*\operatorname{grad}\psi - \psi\operatorname{grad}\psi^*)\end{aligned}\right\} \tag{43}$$

as the densities of electric charge and electric current. We see that the use of a complex ψ is not merely a formal abbreviation. The plane harmonic wave

$$\psi = a\, e^{-i\omega t + i\mathbf{k}\mathbf{x}}$$

gives the charge and current densities

$$\rho \sim \zeta a^* a$$

$$s = \frac{\mathbf{k}}{\sigma}\rho = \mathbf{v}\rho.$$

Waves and Particles

Electrically charged matter acts upon itself, i.e., the potential U (41) is not merely the potential of an external field but, by

$$\Delta U \sim -\zeta \psi^* \psi$$

it depends on the distribution of matter. The physical system is thus described by the equations

$$\left.\begin{aligned}
-\frac{1}{2\sigma}\Delta\psi + \zeta U\psi - i\dot\psi &= 0 \\
-\frac{1}{2\sigma}\Delta\psi^* + \zeta U\psi^* + i\dot\psi^* &= 0 \\
\varepsilon_0 \Delta U + \zeta \psi^* \psi &= 0
\end{aligned}\right\} \tag{44}$$

A field imposed externally is represented by the boundary conditions for U. In the limiting case of sufficiently thinly distributed matter the last of the equations (44) may be omitted and U may be regarded as the potential due to an external field.

In reality, however, there are such things as particles of matter. The relationships

$$E = \hbar\omega \quad p = \hbar k \quad m = \hbar\sigma \quad \pm e = \hbar\zeta \tag{45}$$

exist between the particle properties E, p, m, charge $\pm e$ and the wave properties ω, k, σ, ζ. If we express σ and ζ in terms of m and e we have the equations

$$\left.\begin{aligned}
-\frac{\hbar^2}{2m}\Delta\psi + V\psi - i\hbar\dot\psi &= 0 \\
-\frac{\hbar^2}{2m}\Delta\psi^* + V\psi^* + i\hbar\dot\psi^* &= 0
\end{aligned}\right\} \tag{46}$$

for ψ and ψ^*. As a single particle does not act upon itself we may use these equations to describe a single particle system, taking V as the potential energy in an external electric field. This is, of course, a non-intuitive modification of what was previously an intuitive field theory of matter. If we express the fact that we are dealing with precisely one particle by the equation

$$\int \psi^* \psi \, d\tau = 1$$

we may interpret $\psi^*\psi$ as the probability of finding the particle in the region $d\tau = dx\, dy\, dz$. For a constant energy (46) takes the form

$$-\frac{\hbar^2}{2m}\Delta\psi + (V-E)\psi = 0. \tag{36}$$

For a single particle the non-intuitive modifications of classical particle mechanics and the non-intuitive modifications of an intuitive field theory of matter lead to the same quantum equation (36). Particle mechanics is thus modified by replacing the variables with operators which satisfy the commutation relations; in particular we replace p_k by $\hbar\partial/i\partial q_k$. The classical equation

$$\frac{p^2}{2m} + V(x,y,z) - E = 0$$

thus led to equation (36). Intuitive field theory was modified by assuming the existence of a particle which did not act upon itself. The intuitive equation (42) became equation (36) by putting $m = \hbar\sigma$, $\pm e = \hbar\zeta$, $E = \hbar\omega$, and $\psi^*\psi\, d\tau$ was interpreted as a probability.

The Schroedinger Equation

The non-intuitive modification of an intuitive field theory of matter into the quantum mechanics of a many-particle system had to be carried out in the context of equations (44), which was not easy. The modification of classical particle mechanics is, however, independent of the number of degrees of freedom. The equation

$$H(p_1, p_2 \ldots q_1, q_2 \ldots) - E = 0$$

gave the SCHROEDINGER equation

$$\left\{ H\left(\frac{\hbar}{i}\frac{\partial}{\partial q_1}, \frac{\hbar}{i}\frac{\partial}{\partial q_2} \ldots q_1, q_2 \ldots\right) - i\hbar\frac{\partial}{\partial t} \right\} \psi(q_1, q_2 \ldots t) = 0. \tag{47}$$

If we put

$$\int \psi^*\psi\, d\tau = 1$$

where the integral is taken over the space of all the co-ordinates $x_1, y_1, z_1, x_2, \ldots, z_N$, $\psi^*\psi\, d\tau$ denotes the probability that the co-ordinates of the N particles lie in the region $d\tau = dx_1 \ldots dz_N$, for $x_1, y_1, \ldots, z_N$.

A function

$$\psi = u(q_1, q_2 \ldots) e^{-i\omega t}$$

for which

$$H\psi = i\hbar\dot{\psi} = \hbar\omega\psi$$

denotes the state of definite energy $E = \hbar\omega$. A state of a conservative system with indeterminate energy can be represented as

$$\psi = \Sigma c_n u_n(q_1, q_2 \ldots) e^{-i\omega_n t}.$$

$c_n^* c_n$ is the probability that the system has energy $E_n = \hbar\omega_n$ (if $\int u_n^* u_n \, d\tau = 1$). In the case of the single particle system the function

$$\psi \sim e^{ikx}$$

for which

$$\frac{\hbar}{i} \frac{\partial\psi}{\partial x} = \hbar k\psi$$

denotes a state with the definite value $p_x = \hbar k$ of the x-components of the momentum;

$$\psi \sim e^{i\mathbf{k}x}$$

with

$$\frac{\hbar}{i} \operatorname{grad}\psi = \hbar\mathbf{k}\psi$$

denotes a state with definite momentum vector $\mathbf{p} = \hbar\mathbf{k}$. A state with indeterminate momentum can be represented by

$$\psi(x, y, z) = \frac{1}{\sqrt{2\pi}^3} \quad \varphi(p_x, p_y, p_z) e^{\frac{i}{\hbar}\mathbf{p}x} dp_x dp_y dp_z.$$

A 'box potential', $V = 0$ for $0 < x < a$ and $V = \infty$ otherwise, has the eigenfunctions

$$u_n = \sqrt{\frac{2}{a}} \sin kx = \frac{1}{i}\sqrt{\frac{1}{2a}} (e^{ikx} - e^{-ikx}) \qquad k = \frac{\pi n}{a}$$

and thus a definite value of $p^2 = \hbar^2 k^2$, while the values $\pm\hbar k$ of p occur with equal probability. For a particle the angular momentum operator P_z about the z direction is

$$\frac{\hbar}{i}\left(x\frac{\partial}{\partial y} - y\frac{\partial}{\partial x}\right) = \frac{\hbar}{i}\frac{\partial}{\partial\varphi}$$

where φ is the azimuthal angle about the z axis. A function

$$\varphi \sim e^{im\varphi}$$

such that

$$\frac{\hbar}{i}\frac{\partial\psi}{\partial\varphi} = m\psi$$

thus denotes a state with a definite value of the angular momentum $P_z = \hbar m$. The equation

$$i(P_x P_y - P_y P_x) = \hbar P_z$$

which is easily derived, for the operators corresponding to the angular momentum components P_x, P_y and P_z, shows that no two components of angular momentum may take simultaneous exact values. The surface spherical harmonics $Y_l(\vartheta,\varphi)$ of order l satisfy the equation

$$-\hbar^2\left[\left(x\frac{\partial}{\partial y} - y\frac{\partial}{\partial x}\right)^2 + \left(y\frac{\partial}{\partial z} - z\frac{\partial}{\partial y}\right)^2 + \left(z\frac{\partial}{\partial x} - x\frac{\partial}{\partial z}\right)^2\right]Y_l = l(l+1)Y_l.$$

A function

$$\psi \sim Y_l(\vartheta,\varphi)$$

thus denotes a state with definite value $l(l+1)$ of the square P^2 of the angular momentum vector.

Three Examples

The most important ways of satisfying the commutation relations are the use of matrices, the use of the operator $\hbar d/i\,dq$ for p, and possibly also the use of the operator $i\hbar\,d/dp$ for q. We shall consider the simplest example in each case.

Matrices are easy to handle only in the case of the harmonic oscillator

$$H = \frac{p^2}{2m} + \frac{m\omega^2 x^2}{2}.$$

We may simplify HEISENBERG's own calculation (Chapter 10) slightly. We shall use complex variables, often useful for harmonic oscillations, which here make the connection between the commutation relation and the eigenvalues quite clear. They were destined to become very significant in quantum field theory.

A sum of two squares may be written in the form $\xi^*\xi$. In order to emphasize the essential points we shall put $\hbar$, m and ω all equal to 1 and thus write

$$H = \frac{1}{2}(p^2 + q^2).$$

Using the transformation

$$\frac{1}{\sqrt{2}}(q + ip) = \xi \qquad q = \frac{1}{\sqrt{2}}(\xi + \xi^*)$$

$$\frac{1}{\sqrt{2}}(q - ip) = \xi^* \qquad p = \frac{1}{i\sqrt{2}}(\xi - \xi^*)$$

we obtain

$$H = \tfrac{1}{2}(\xi^*\xi + \xi\xi^*). \tag{48}$$

The equations of motion $\dot{q} = p$, $\dot{p} = -q$ are then

$$\dot{\xi} = -i\xi \qquad \dot{\xi}^* = i\xi^*.$$

These are the canonical equations to (48) with ξ and $i\xi^*$ as canonical conjugate variables. By the commutation relation (33) we obtain

$$\xi\xi^* - \xi^*\xi = 1. \tag{49}$$

The equations of motion show that only the frequency -1 occurs in the matrix ξ and only the frequency $+1$ occurs in ξ^*. The matrices are of the form

$$\xi = \begin{Bmatrix} 0 & \xi_{01} & 0 & \dots \\ 0 & 0 & \xi_{12} & \dots \\ 0 & 0 & 0 & \dots \\ . & . & . & \dots \end{Bmatrix} e^{-it} \qquad \xi^* = \begin{Bmatrix} 0 & 0 & 0 & \dots \\ \xi_{01}^* & 0 & 0 & \dots \\ 0 & \xi_{12}^* & 0 & \dots \\ . & . & . & \dots \end{Bmatrix} e^{it}$$

For $\xi\xi^*$ and $\xi^*\xi$ we obtain diagonal matrices:

$$(\xi\xi^*)_{nn} = |\xi_{n,n+1}|^2$$

$$(\xi^*\xi)_{nn} = |\xi_{n-1,n}|^2$$

if we put $\xi_{-1,0} = 0$. The commutation relation (49) gives

$$(\xi\xi^*)_{nn} = n + 1$$

$$(\xi^*\xi)_{nn} = n.$$

The diagonal values of $\xi^*\xi$ are thus the possible numbers of wave quanta. The energy (W) is a diagonal matrix with the elements

$$E_n = n + \tfrac{1}{2}.$$

If we put in the factors $\hbar$, m and ω, we get $E_n = \hbar\omega(n + \tfrac{1}{2})$.

So far we have allowed the rows and columns of our matrices to correspond to the possible energies of the oscillator. We may, however, use (49) to calculate the eigenvalues of $\xi^*\xi$ directly. If we multiply (49) on the right by ξ, we obtain

$$\xi(\xi^*\xi) - (\xi^*\xi+1)\xi = 0$$

and if $\xi^*\xi$ is to be a diagonal matrix

$$\xi_{nl}[(\xi^*\xi)_{ll} - (\xi^*\xi)_{nn} - 1] = 0$$

in other words, the element ξ_{nl} is zero only if the diagonal elements of $\xi^*\xi$ which correspond to l and n differ by 1. The rows and columns can be arranged in such a way that $l = n + 1$ and $(\xi^*\xi)_{nn} = n$, where we have used the fact that $\xi_{-1,0} = 0$. If the matrices ξ and ξ^* satisfy (49) and if $\xi^*\xi$ is a diagonal matrix, the diagonal consists of the numbers $0, 1, 2, \ldots$

It is possible to multiply a column on the left by a matrix. If we represent a state in which there are n wave quanta by the column vector

$$\Phi_n = \begin{pmatrix} 0 \\ 0 \\ \vdots \\ 1 \\ 0 \\ \vdots \end{pmatrix}$$

which has a 1 in the n-th row, the multiplication gives

$$\xi\Phi_n = \sqrt{n}\,\Phi_{n-1}$$
$$\xi^*\Phi_n = \sqrt{n+1}\,\Phi_{n+1}.$$

The 'operator' ξ^* increases the number of wave quanta by 1, the operator ξ decreases it by 1. ξ applied to the ground state Φ_0 gives zero, while ξ^* applied to the ground state Φ_0 gives Φ_1. We describe ξ^* and ξ as creation and annihilation operators. They play an important role in quantum field theory, where particles are treated as wave quanta.

The use of the operator $\hbar d/i\, dq$ for p leads with the kinetic energy $p^2/2m$ to the simple differential equation

$$-\frac{\hbar^2}{2m}\psi'' + [V(q) - E]\psi = 0 \tag{50}$$

which often represents an eigenvalue problem. For a $V(q)$ which increases in both directions we see immediately that a function ψ without zeros corresponds to the lowest (zero) eigenvalue, and that a function with n zeros corresponds to the n-th eigenvalues. The simplest examples are the 'box'

$$q = x \qquad V(x) = \begin{cases} 0 & 0 < x < a \\ \infty & x < 0, x > a \end{cases}$$

and the free rotator

$$q = \varphi \qquad V(\varphi) = 0.$$

For the box, ψ must satisfy the boundary condition $\psi = 0$ at 0 and a. (50) can be solved by means of

$$\psi \sim \sin\frac{\pi n x}{a} \qquad E = \frac{\pi^2 \hbar^2 n^2}{2ma^2}.$$

In the case of the rotator ψ must be unique, so that:

$$\psi \sim \frac{\cos}{\sin}n\varphi \qquad E = \frac{\hbar^2 n^2}{2I}$$

The operator $i\hbar\, d/dp$ for q is useful only if $V(q)$ is a simple operator. This is the case for $V = -Z/q$ (an example with which we dealt in the context of the correspondence principle).[1] The operator reciprocal to q, $1/q$, can be replaced by the integral operator $\int dp$ (we have left out the irrelevant consideration of the integration constant). The 'SCHROEDINGER equation in momentum space' thus becomes:

$$\frac{iZ}{\hbar}\int\varphi\, dp + \left(\frac{p^2}{2m} - E\right)\varphi = 0$$

and thus, using the abbreviation $-2mE = \eta^2$, as we are interested only in negative values of E, this becomes

$$\frac{\varphi}{\int\varphi\, dp} = -\frac{2iZm}{\hbar(p^2 + \eta^2)}.$$

If we integrate this we get

$$\int \varphi \, dp = e^{-\frac{2iZm}{\hbar\eta} \arctan \frac{p}{\eta}}.$$

This must be a single-valued function of p. The exponent must therefore be $-2in \arctan (p/\eta)$. It follows that:

$$\eta = \frac{Zm}{\hbar n} \qquad E = -\frac{Z^2 m}{2\hbar^2 n^2}.$$

The Formalism

In classical 'kinematics' each state of a system is represented by a point in the space of the canonical variables p_k, q_k. The transition to other canonical variables is effected independently of the dynamics by canonical transformations of the type

$$\bar{p}_k = \bar{p}_k(p_1 \cdots q_1 \cdots) \qquad \bar{q}_k = \bar{q}_k(p_1 \cdots q_1 \cdots).$$

The dynamics of a particular classical system is characterized by the Hamiltonian $H(p_1, \ldots, q_1, \ldots, t)$. The motion satisfies the canonical equations

$$q_k = \frac{\partial H}{\partial p_k} \qquad p_k = -\frac{\partial H}{\partial q_k}. \qquad (51)$$

In quantum mechanics the canonical space is not an ordinary space. Its structure is described by the commutation relations (34), *inter alia* by

$$i(p_k q_l - q_l p_k) = \hbar \delta_{kl}.$$

The (real) observables are represented by Hermitian operators (Hermitian because $A^* = A$) applied to normalized vectors ψ_n (normalized because $\Sigma \psi_n^* \psi_n = 1$). The canonical transformations are 'unitary transformations' of the vectors:

$$\bar{\psi} = S\psi \qquad S^* S = 1$$

to which correspond the transformations

$$\bar{A} = S A S^*$$

of the observables. In quantum dynamics, equations (51) are preserved. By (34) they may be replaced by

$$\dot{A} = \frac{i}{\hbar}(HA - AH)$$

for all observables A.

The vectors ψ_n in quantum mechanics are analogous to the vectors of an ordinary space, for which the transformation to another 'basis' is effected by means of the transformation

$$\bar{\psi}_n = \sum_m S_{nm}\psi_m \qquad (\bar{\psi} = S\psi).$$

If we restrict ourselves to orthogonal bases we have

$$\sum_n S_{nk} S_{nl} = \delta_{kl}$$

$$\sum_n \tilde{S}_{kn} S_{nl} = \delta_{kl} \qquad (\tilde{S}S = 1).$$

Apart from the orthogonal changes of basis S which are analogous to the canonical transformations, we shall consider representations A. These are in general composed of dilation and rotation:

$$\varphi_n = \sum_m A_{nm}\psi_m \qquad (\varphi = A\psi).$$

Using another basis we have $\varphi = S^{-1}\bar{\varphi}$ and $\psi = S^{-1}\bar{\psi}$, so that

$$\bar{\varphi} = SAS^{-1}\bar{\psi}$$

$$\bar{A} = SAS^{-1}.$$

For an orthogonal change of basis we have

$$\bar{A} = SA\tilde{S}.$$

The symmetrical representations

$$A_{nm} = A_{mn} \qquad A = \tilde{A}$$

are particularly important, being pure dilations. They can be transformed into principal axes by means of an orthogonal transformation. The eigenvectors ψ_a, such that

$$A\psi_a = a\psi_a$$

with eigenvalues a are, or can be chosen in such a way that they are, orthogonal if there is more than one eigenvector corresponding to a single eigenvalue. The symmetrical representations correspond to the observables.

As we also wish to be able to use non-real vectors ψ_n in quantum mechanics, let us generalize the vector space with the invariants $\Sigma\psi_n\psi_n$; we then have the 'unitary basis transformations' with the invariant $\Sigma\psi_n^*\psi_n$. If we use the convention

$$(S^*)_{nm} = S_{mn}^*$$

a unitary basis transformation satisfies

$$S^* S = 1 \tag{51}$$

and a representation A is transformed by the unitary basis transformation S into

$$\bar{A} = S A S^*. \tag{52}$$

In place of the symmetrical representations we now have the Hermitian representations

$$A^* = A.$$

These can be transformed into principal axes by a unitary basis transformation. Their eigenvalues are real.

In quantum mechanics the vector space has an infinite number of dimensions, and the indices of the vector components may be continuous. The changes of basis correspond to changes in 'aspect'.

In the 'position aspect' of the SCHROEDINGER equation, x corresponds to the index, $\psi(x)$ is the state vector. In the momentum aspect p corresponds to the index and the state vector is called $\psi(p)$. The most important bases or 'aspects' in quantum theory are those of position, momentum and 'n', which may be, for example, an energy. Table 9 is designed to remind us of the appropriate matrices or operators S.

TABLE 9: CHANGES OF BASIS

	x	p	n
x	$\psi(x) = \int \psi(x')\delta(x-x')\mathrm{d}x'$	$\psi(x) = \dfrac{1}{\sqrt{2\pi}} \int \varphi(p)\mathrm{e}^{\frac{\mathrm{i}}{\hbar}px}\mathrm{d}p$	$\psi(x) = \sum_n a_n u_n(x)$
p	$\varphi(p) = \dfrac{1}{\sqrt{2\pi}} \int \psi(x)\mathrm{e}^{-\frac{\mathrm{i}}{\hbar}px}\mathrm{d}x$	$\varphi(p) = \int \varphi(p')\delta(p-p')\mathrm{d}p'$	$\varphi(p) = \sum_n a_n v_n(p)$
n	$a_n = \int \psi(x)u_n^*(x)\mathrm{d}x$	$a_n = \int \varphi(p)v_n^*(p)\mathrm{d}p$	$a_n = \sum_n a_m \delta_{mn}$

The physical quantities, the observables, correspond to Hermitian operators. Their eigenvalues give the possible values of the observables, as measured. The transition magnitude $|S_{nk}|^2$ gives the probability of finding for a fixed value n in the n basis the value k of the k basis. With the notation of the table $|u_n(x)|^2$

is the probability, in a state which corresponds to the index n, of finding the position in the interval $(x, x + \mathrm{d}x)$.

Let us choose a very simple example, a physical system in which a particle can be in one of two positions, let us say left or right. The basis vectors in this position representation will be called l and r.

$$\psi = \psi_l l + \psi_r r \tag{53}$$

is then an arbitrary state vector. Instead of the position representation, we may take a symmetry representation with the vectors

$$a = \frac{1}{\sqrt{2}}(l-r) \qquad s = \frac{1}{\sqrt{2}}(l+r)$$

as basis. The state vector is then

$$\psi = \psi_s s + \psi_a a. \tag{54}$$

Comparison of (53) and (54) gives the transformation

$$\begin{pmatrix} \psi_l \\ \psi_r \end{pmatrix} = \begin{pmatrix} \dfrac{1}{\sqrt{2}} & \dfrac{1}{\sqrt{2}} \\ -\dfrac{1}{\sqrt{2}} & \dfrac{1}{\sqrt{2}} \end{pmatrix} \begin{pmatrix} \psi_a \\ \psi_s \end{pmatrix}$$

The probability of finding the position l and r in the state s or a is $\frac{1}{2}$ in each case. As the vectors are here all real we may represent the transformation as in Figure 33.

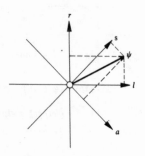

FIGURE 33: SIMPLE TRANSFORMATION OF BASIS

[1] G. Rumer, C. R. Akad. USSR **3**, 102 (1933)

CENTRES OF RESEARCH

Berlin[1]

IN the decades around the turn of the century Berlin was one of the foremost research centres in the world. From 1871 to 1888 HELMHOLTZ had directed the Physical Institute at the university and afterwards the Physikalisch-Technische Reichsanstalt. KIRCHHOFF had occupied the chair in theoretical physics at the university from 1874 to 1888. His successor was PLANCK. Around 1880 or shortly thereafter PLANCK (for a short while), the spectroscopists KAYSER, RUNGE, PRINGSHEIM, LUMMER, KURLBAUM and RUBENS, not to mention WIEN and HABER were studying in Berlin. The measurements of series spectra by KAYSER and RUNGE were, however, first carried out in Hanover. The crucial measurements were made at the Physikalisch-Technische Reichsanstalt. WIEN was there when he made his discoveries of the displacement law and his radiation formula. Measurements made by LUMMER and PRINGSHEIM made PLANCK aware of black-body radiation. The deviations from WIEN's formula that these two men discovered, and perhaps more particularly the precise measurements made by RUBENS and KURLBAUM, created the experimental requirements for PLANCK's formula.

In the years before the First World War theoretical physics was represented by PLANCK, VON LAUE and EINSTEIN (from 1914), and experimental physics by RUBENS and KURLBAUM. NERNST and HABER directed institutes whose work was very close to physics. Berlin at that time attracted a number of younger workers: FRANCK, POHL, HERTZ, PRINGSHEIM, GEIGER, BOTHE, HAHN, and MEITNER. The 'Wednesday Colloquium' had only a small number of participants but it was widely renowned, not least as a result of the lively discussions with EINSTEIN about the relationship between light quanta and light waves and on the theory of relativity.

Berlin continued to attract young workers in the twenties: GORDON, LONDON and above all the 'extraordinarily intelligent Hungarians' VON NEUMANN, WIGNER and SZILARD. In 1927 SCHROEDINGER was summoned to Berlin to succeed PLANCK.

Cambridge, England

Although in England scientific research was carried out by wealthy individuals to a far greater extent than on the Continent, there were important research centres at many of the universities. In physics the most important must surely have been Cambridge, where the Cavendish laboratory was the home of much important work. It was directed successively by MAXWELL, RAYLEIGH, for more than thirty years by J. J. THOMSON, and then by RUTHERFORD, who had fathered nuclear physics while in Manchester. Around 1925, an important year for quantum mechanics, R. H. FOWLER was pre-eminent in theoretical physics at Cambridge. P. A. M. DIRAC was just completing his studies under FOWLER. We must not forget LARMOR, nor EDDINGTON, who was very interested in physics, nor of course old J. J. THOMSON, who all formed part of the Cambridge scene. We should also include C. G. DARWIN, who was in Cambridge before and after this period and who should be counted as one of the circle.

Cambridge attracted visiting scientists: LANDAU discovered his theory of the diamagnetism of free electrons in Cambridge and NORDHEIM used his time there to write about the electron theory of metals.

Cambridge, Massachusetts

In the twenties Harvard University possessed an important spectroscopic research centre, the fame of which was due mainly to LYMAN. H. N. RUSSELL and F. A. SAUNDERS discovered the vector model of the atom with two external electrons. At about the same time DUANE came very close to founding wave mechanics. Among the physics students who learned quantum theory from E. C. KEMBLE were R. S. MULLIKEN, J. C. SLATER and J. H. VAN VLECK. MULLIKEN soon applied it to molecules, SLATER to rigid bodies, and VAN VLECK used it to tackle fundamental problems.

Göttingen[2]

From the time of GAUSS, Göttingen had been one of the fortresses of mathematics and many of the 'princes of mathematics' were at Göttingen University. From WEBER onwards, physics had also established a fine reputation in Göttingen.

In 1895 HILBERT came to Göttingen to join Felix KLEIN, who was somewhat older. SOMMERFELD was working with KLEIN at that time, but he soon went to Clausthal and then to Aachen. VOIGT taught theoretical physics and NERNST physical chemistry. It was in Göttingen that NERNST discovered his heat theorem. A large number of foreigners came to study in Göttingen; for example, MILLIKAN and LYMAN as well as H. M. HANSEN who was responsible for acquainting BOHR with the spectral laws. Apart from them, RITZ worked intermittently in Göttingen, VON LAUE was there for a few years, and STARK began his lively scientific career here. An increase in the number of chairs brought WIECHERT, PRANDTL and RUNGE to Göttingen in 1905. All three were keenly interested in the fundamental problems of physics. 1904 saw the arrival of BORN as a student of mathematics. He worked closely with HILBERT and MINKOWSKI. In the years before the First World War there was a very lively circle of young scientists, consisting of BORN, WEYL, VON KARMAN, COURANT and EWALD.

At a conference in 1913, the young DEBYE impressed the Göttingen scientists, and it was mostly due to the efforts of HILBERT that he was called to Göttingen as a professor (1914–20). He was followed by BORN and FRANCK. Quantum theory was still in its early stages, and it made a quick upturn under their influence, while POHL laid the foundations for important areas of rigid body physics. It was in Göttingen in 1922 that BOHR gave his famous lectures. It was in Göttingen that the matrix formulation and the probabilistic interpretation of quantum mechanics were developed. In the twenties the following people were working in BORN's institute: PAULI, HUECKEL, LONDON, HEISENBERG, FERMI, JORDAN, HUND, NORDHEIM, HEITLER, E. A. HYLLERAS, FRENKEL, KRONIG, FOCK, J. R. OPPENHEIMER, WIGNER, ROSENFELD, Maria GÖPPERT, DELBRÜCK and WEISSKOPF.

Copenhagen

Niels BOHR made Copenhagen one of the main centres of quantum theory. He studied in Copenhagen and then worked with Ernest RUTHERFORD in Manchester. In 1916 he became professor of theoretical physics at Copenhagen. Among his pupils at the beginning were H. A. KRAMERS and Oskar KLEIN. The Universitetets Institut for Teoretisk Fysik, founded in 1920 and renamed the Niels Bohr Institute in 1965 soon became the spiritual home of a closely knit group of physicists. KRAMERS and later HEISENBERG and KLEIN taught at the institute. Among the guests who came in the twenties for varying lengths of time were: FRANCK, RUBINOWICZ, VON HEVESY, PAULI, VAN VLECK, NISHINA, SLATER, KUHN, FOWLER, THOMAS, JORDAN, DENNISON, GOUDSMIT, HUND, FUES, HEITLER, PAULING, KRONIG, NORDHEIM, HARTREE, GAMOW, LANDAU and CASIMIR. After 1933 the BOHR Institute became a sanctuary for refugee physicists.

Most of BOHR's early papers were written in Copenhagen. The logical development of the correspondence principle from dispersion theory right up to the matrix formulation, the clarification of basic concepts, and the connection with semantics are all part of the Copenhagen spirit. It is impossible to assess the stimulating effect on the course of physics and philosophy of the ideas that emanated from BOHR's institute.

Leiden and Utrecht

Holland has produced a relatively large number of major physicists, and ever since the seventeenth century Leiden has been a centre of scientific research. Around the turn of the last century Hendrik Antoon LORENTZ was active there, and L. S. ORNSTEIN was one of his pupils. In 1912 LORENTZ was succeeded by EHRENFEST, who had already made a contribution to quantum theory in St Petersburg who went on to establish his adiabatic hypothesis in Leiden. Among his pupils were: BURGERS, KRAMERS, COSTER, GOUDSMIT, UHLENBECK and CASIMIR. One of his collaborators was BREIT. A number of branches of physics were stimulated by work carried out at the Leiden Refrigeration Institute. The liquefaction of helium gas which was carried out in 1908 facilitated the measurements of specific heats at very low temperatures which were vital for early quantum theory.

It was in Utrecht that ORNSTEIN and his colleagues established the exact intensity laws for spectral lines in 1924. These were a preliminary to the development of a rigorous form of quantum mechanics. In 1925 KRAMERS became a professor of theoretical physics in Utrecht.

Leipzig[3]

When in 1927 the occupants of the chairs of theoretical and experimental physics both died within a short time, the faculty sent for DEBYE and HEISENBERG. They were soon joined by the mathematician B. L. VAN DER WAERDEN and later by the physio-chemist K. F. BONHOEFFER.

The brief blossoming of Leipzig physics began when the foundations of quantum theory had already been laid. But important applications were developed in Leipzig, including the theory of ferromagnetism and metal electrons (HEISENBERG and BLOCH), and contributions to quantum chemistry (HUECKEL). To a large extent the basis of quantum electrodynamics and of quantum field theory and also theoretical nuclear physics all occurred in Leipzig. Around 1930 HEISENBERG's and DEBYE's institutes drew important visitors from all over the world and the 'Leipzig weeks' on current problems of physics were major events.

Munich

The importance of Munich in the development of quantum theory rests on the achievement of SOMMERFELD[4] as a scientist and as a remarkable magnet to other scientists. After successful work in various fields of theoretical physics (gyroscopic theory, diffraction of waves, technical problems, hydrodynamics, and relativity theory) he was the driving force in German theoretical research into the structure of the atom in the decisive years before and after 1920. As perhaps no other man has ever been able to do, he could grip the listeners to his lectures, convey his own enthusiasm to them and give his pupils tasks which helped them to develop as physicists. Of all the German teachers of theoretical physics he surely had the largest number of important pupils. In 1906 he came to Munich as professor and brought DEBYE with him as his assistant. Among SOMMERFELD's pupils and collabora-tors were: LANDÉ, EPSTEIN, LENZ, EWALD, VON LAUE (the inter-ference of X-rays was discovered in Munich), KOSSEL, RUBI-NOWICZ, HERZFELD, FUES, KRATZER, WENTZEL, PAULI, HEISEN-BERG, CATALAN, BECHERT, PEIERLS, HEITLER, UNSÖLD, BETHE and HOUSTON.

It was in Munich that quantum theory was extended to cover multiply-periodic systems, and a systematic theory was developed, particularly the n, l, j method for spectra, a large part of the understanding of multiplets and later the application of quantum to electrons in metals (SOMMERFELD and BETHE).

Zürich

EINSTEIN, SCHROEDINGER and PAULI[5] were responsible for Zürich's significance in the history of quantum theory. EINSTEIN studied in Zürich and was later professor first at the university (1909–11) and then at the Technische Hochschule (1912–3). His most important early contributions to quantum theory (like those to the theory of relativity) were done while he was an employee of the patent office in Berne. At the university he was succeeded by DEBYE, VON LAUE, SCHROEDINGER (1921–7) and WENTZEL. At the Technische Hochschule, the larger of the two scientific institutes in Zürich, the professors were WEYL, DEBYE (1920–7) and PAULI (1928–58). Among PAULI's colleagues were: KRONIG, PEIERLS, ROSENFELD, CASIMIR and WEISSKOPF. It was in Zürich that the SCHROEDINGER equation was discovered and that PAULI made his contributions to quantum field theory.

[1] W. WESTPHAL in the Gedenkschrift zur 150. Wiederkehr des Gründungsjahres der Universität Berlin, Berlin 1960
 H. KALLMANN, Phys. Bl. **22**, 489 (1966)
[2] E. A. HYLLERAAS, Rev. mod. phys. **35**, 421 (1963)
 F. HUND in the Heisenberg-Festschrift, Braunschweig 1961
[3] F. HUND loc. cit.
[4] M. BORN, A. J. W. SOMMERFELD, Obit. Notices Roy. Soc. **8**, 275 (1952)
[5] P. JORDAN, Begegnungen, Oldenburg 1971; Phys. B6 **29**, 291 (1973)

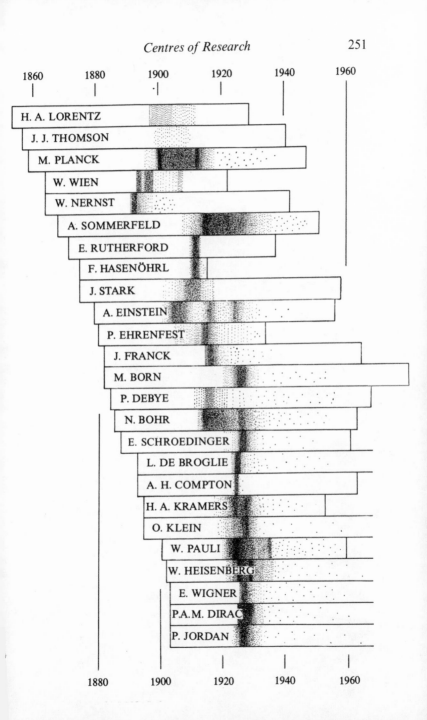

INDEX OF NAMES

SUBJECT INDEX